Application of Nanoparticles for Oil Recovery

Application of Nanoparticles for Oil Recovery

Editor

Ole Torsaeter

MDPI • Basel • Beijing • Wuhan • Barcelona • Belgrade • Manchester • Tokyo • Cluj • Tianjin

Editor
Ole Torsaeter
Department of Geoscience and
Petroleum
Norwegian University of Science
and Technology
Trondheim
Norway

Editorial Office
MDPI
St. Alban-Anlage 66
4052 Basel, Switzerland

This is a reprint of articles from the Special Issue published online in the open access journal *Nanomaterials* (ISSN 2079-4991) (available at: www.mdpi.com/journal/nanomaterials/special_issues/oil_recover).

For citation purposes, cite each article independently as indicated on the article page online and as indicated below:

LastName, A.A.; LastName, B.B.; LastName, C.C. Article Title. *Journal Name* **Year**, *Volume Number*, Page Range.

ISBN 978-3-0365-1318-8 (Hbk)
ISBN 978-3-0365-1317-1 (PDF)

Contents

Preface to "Application of Nanoparticles for Oil Recovery" . . . . **vii**

Ole Torsæter
Application of Nanoparticles for Oil Recovery
Reprinted from: *Nanomaterials* **2021**, *11*, 1063, doi:10.3390/nano11051063 . . . . **1**

Alberto Bila and Ole Torsæter
Experimental Investigation of Polymer-Coated Silica Nanoparticles for EOR under Harsh
Reservoir Conditions of High Temperature and Salinity
Reprinted from: *Nanomaterials* **2021**, *11*, 765, doi:10.3390/nano11030765 . . . . **5**

Nanji J. Hadia, Yeap Hung Ng, Ludger Paul Stubbs and Ole Torsæter
High Salinity and High Temperature Stable Colloidal Silica Nanoparticles with Wettability
Alteration Ability for EOR Applications
Reprinted from: *Nanomaterials* **2021**, *11*, 707, doi:10.3390/nano11030707 . . . . **23**

**Zachary Paul Alcorn, Tore Føyen, Jarand Gauteplass, Benyamine Benali, Aleksandra Soyke
and Martin Fernø**
Pore- and Core-Scale Insights of Nanoparticle-Stabilized Foam for CO_2-Enhanced Oil Recovery
Reprinted from: *Nanomaterials* **2020**, *10*, 1917, doi:10.3390/nano10101917 . . . . **37**

Shidong Li, Yeap Hung Ng, Hon Chung Lau, Ole Torsæter and Ludger P. Stubbs
Experimental Investigation of Stability of Silica Nanoparticles at Reservoir Conditions for
Enhanced Oil-Recovery Applications
Reprinted from: *Nanomaterials* **2020**, *10*, 1522, doi:10.3390/nano10081522 . . . . **53**

Edgar Rueda, Salem Akarri, Ole Torsæter and Rosangela B.Z.L. Moreno
Experimental Investigation of the Effect of Adding Nanoparticles to Polymer Flooding in
Water-Wet Micromodels
Reprinted from: *Nanomaterials* **2020**, *10*, 1489, doi:10.3390/nano10081489 . . . . **69**

Reidun C. Aadland, Salem Akarri, Ellinor B. Heggset, Kristin Syverud and Ole Torsæter
A Core Flood and Microfluidics Investigation of Nanocellulose as a Chemical Additive to Water
Flooding for EOR
Reprinted from: *Nanomaterials* **2020**, *10*, 1296, doi:10.3390/nano10071296 . . . . **91**

**Amin Rezaei, Hadi Abdollahi, Zeinab Derikvand, Abdolhossein Hemmati-Sarapardeh,
Amir Mosavi and Narjes Nabipour**
Insights into the Effects of Pore Size Distribution on the Flowing Behavior of Carbonate Rocks:
Linking a Nano-Based Enhanced Oil Recovery Method to Rock Typing
Reprinted from: *Nanomaterials* **2020**, *10*, 972, doi:10.3390/nano10050972 . . . . **115**

Preface to "Application of Nanoparticles for Oil Recovery"

The oil industry has, in the last decade, seen successful applications of nanotechnology in completion systems, completion fluids, drilling fluids, and in improvements of well constructions, equipment, and procedures. However, very few full field applications of nanoparticles as an additive to injection fluids for enhanced oil recovery (EOR) have been reported. Many types of chemical enhanced oil recovery methods have been used in fields all over the world for many decades and have resulted in higher recovery, but the projects have very often not been economic. Therefore, the oil industry is searching for a more efficient enhanced oil recovery method. Based on the success of nanotechnology in various areas of the oil industry, nanoparticles have been extensively studied as an additive in injection fluids for EOR. This book includes a selection of research articles on the use of nanoparticles for EOR application. The articles are discussing nanoparticles as additive in waterflooding and surfactant flooding, stability and wettability alteration ability of nanoparticles and nanoparticle stabilized foam for CO2-EOR. The book also includes articles on nanoparticles as an additive in biopolymer flooding and studies on the use of nanocellulose as a method to increase the viscosity of injection water. Mathematical models of the injection of nanoparticle-polymer solutions are also presented.

Ole Torsaeter

Editor

nanomaterials

Editorial

Application of Nanoparticles for Oil Recovery

Ole Torsæter

Department of Geoscience and Petroleum, Norwegian University of Science and Technology (NTNU), 7031 Trondheim, Norway; ole.torsater@ntnu.no

Citation: Torsæter, O. Application of Nanoparticles for Oil Recovery. *Nanomaterials* **2021**, *11*, 1063. https://doi.org/10.3390/nano11051063

Received: 7 April 2021
Accepted: 20 April 2021
Published: 21 April 2021

Publisher's Note: MDPI stays neutral with regard to jurisdictional claims in published maps and institutional affiliations.

Due to their large surface-area-to-volume ratio and enhanced chemical reactivity, nanoparticles have attracted interest among researchers in the upstream petroleum industry for oil recovery applications. Nanoparticles have been studied as additives to waterflooding from day one of production as well as additives at later-stage waterflooding (secondary and tertiary recoveries). Many types of nanoparticles have been tested, and the aims of the nanoparticles have been either to reduce the mobility of the injected fluid relative to that of oil or to increase the ratio of viscous to interfacial forces (i.e., capillary number).

The research on nanotechnology for oil recovery has shown potential, but the mechanisms for oil recovery are not fully understood. This Special Issue of *Nanomaterials*, "Application of Nanoparticles for Oil Recovery", includes eight research articles that demonstrate advances in the understanding of the behavior of nanoparticles during fluid flow in porous media and give more knowledge of recovery mechanisms when using nanoparticles in oil recovery. The research works presented here cover core flooding and microfluidic studies of the injection of various nanoparticle suspensions (in water and together with other chemicals) for enhanced oil recovery (EOR), nanostabilized CO_2 foam flooding, investigation of stability of nanosuspensions under various conditions, and mathematical modelling of injection processes of polymer with nanoparticles.

Studies on the use of surface-functionalized particles for special reservoir rock and fluid properties and reservoir conditions might be the way forward for understanding the oil recovery mechanisms involved. An article in this Special Issue by Bila and Torsæter [1] investigated tertiary flooding of polymer-coated silica nanoparticles for oil recovery at high temperature and high salinity. The flooding tests indicated incremental oil recovery of up to 6% of the original oil in place (OOIP). Among the recovery mechanisms studied, the change in rock wettability to more water-wet seemed to be most important.

The stability of nanoparticles is important for a successful application of nanofluids for enhanced oil recovery. Hadia et al. [2] studied the stability of commercial silica nanoparticles in combination with zwitterionic and hydrophilic silanes under high salinity and high temperature conditions. The results showed thermal stability in synthetic seawater at 60 °C for 1 month and accelerated stability analysis predicted that the modified nanoparticles could remain stable for at least 6 months. The ability of the nanoparticles to alter rock wettability was also studied. The results showed that the modified nanoparticles were able to adsorb on rock surfaces and altered wettability to water-wet.

Previous research has shown that nanoparticles can increase the stability of surfactant-based foams. Stable foams are especially important in processes where CO_2 foam is used to recover oil and in CO_2 storage. The behavior of nanoparticle–surfactant foam formulations is not well understood, and the experimental work by Alcorn et al. [3] presents a pore-to-core-scale study of foam behavior. Snap-off was the most important foam generation mechanism in high-pressure micromodels. It was also observed that foaming solutions containing only nanoparticles generated very little foam. When nanoparticles and surfactant were used together, foam generation and strength were not sensitive to the nanoparticle concentration. The experiments with oil showed that the apparent viscosity of foam was nearly three times higher than that in the experiments without oil. This was due to the development of stable oil/water emulsions.

1

Davarpanah [4] presents a mathematical model for simultaneous injection of polymer-assisted nanoparticles for calculation of an oil recovery factor. A sensitivity analysis is provided to show the influence of formation damage in a nanoparticles–polymer solution injection process. The study shows that large mobility ratio, high polymer concentration, and more formation damage lead to increased recovery factor. The reason for this is that the external filter cake is building up in this period and the subsequent injection of the polymer solution thereby gives a better sweep efficiency and higher oil recovery.

The stability of a nanoparticle suspension at reservoir conditions is still creating uncertainty in the evaluation of nanoflooding processes. Li et al. [5] studied the effect of nanoparticle treatment on the stability at reservoir conditions in the presence of reservoir rock and crude oil. The stability was screened in test tubes at 70 °C and 3.8 wt. % NaCl. Fumed silica nanoparticles in suspension with hydrochloric acid (HCl), polymer-modified fumed nanoparticles, and amide-functionalized silica colloidal nanoparticles were studied. The results showed that both HCl and polymer surface modification can improve nanoparticle stability. It was also found that pH is important for nanoparticle stability. Based on the results in this study, a stabilizer and/or nanoparticles modification are necessary for EOR application.

The experimental study by Rueda et al. [6] showed the effect of adding nanoparticles in biopolymer flooding. The tests were performed in water-wet microfluidics setup using xanthan gum and scleroglucan, as well as silica-based nanoparticles in a secondary flooding mode where the capillary number was kept constant. The reference nanofluid flood resulted in higher ultimate oil recovery than similar floods with xanthan gum, scleroglucan, and brine. When adding nanoparticles to the biopolymer solutions, nano-xanthan flooding achieved the highest oil recovery. A reduced polymer adsorption in the nano-xanthan flooding may explain the improvement in the sweep efficiency and recovery factor.

Aadland et al. [7] conducted core flood and microfluidics experiments with fluids containing cellulose nanocrystals (CNCs) and 2,2,6,6-tetramethylpiperidine-1-oxyl (TEMPO)-oxidized cellulose nanofibrils (T-CNFs). Both particles were mixed in brine, and the oil recovery was compared with standard water flooding. CNCs produced 5.8% more of the original oil in place (OOIP) than pure brine flooding. The effect of injection scheme, temperature, and rock wettability was also investigated for both the CNC injection and brine injection, and CNC performed better than pure brine in all cases. The study of T-CNFs showed that these particles were even more effective than CNCs. However, the injectivity gradually deteriorated, and work is ongoing to solve this problem.

Rezaei et al. [8] investigated the effect of the pore throat size distribution on oil recovery by surfactant–nanofluid injection. Interfacial tensions and contact angles were measured to determine the optimum concentrations of an anionic surfactant and silica nanoparticles for core flooding experiments. The results of relative permeability tests showed that the pore throat size distribution affected the endpoints of the relative permeability curves, and a large amount of unswept oil was recovered by the flooding surfactant and silica nanoparticles. The results of the core flooding tests indicated that the injection in tertiary mode increased the post-water flooding oil recovery by up to 2.5% for carbonate core plugs with homogeneous pore throat size distribution and 8.6% for cores with heterogeneous pore throat size distribution.

Funding: This research was funded by PoreLab Center of Excellence, Department of Geoscience and Petroleum, Norwegian University of Science and Technology, grant number 262644.

Acknowledgments: The Guest Editor extends his gratitude to all authors for submitting their work to this Special Issue and for its successful completion. A special recognition is also extended to all the reviewers that contributed to ensuring the quality and impact of the published work. The Guest Editor is also grateful to Greta Zhang and the whole editorial team of Nanomaterials for their constant assistance and support.

Conflicts of Interest: This editorial was written by the coauthor of five of the eight research articles presented in this Special Issue. The author declares no conflict of interest.

References

1. Bila, A.; Torsæter, O. Experimental Investigation of Polymer-Coated Silica Nanoparticles for EOR under Harsh Reservoir Conditions of High Temperature and Salinity. *Nanomaterials* **2021**, *11*, 765. [CrossRef]
2. Hadia, N.J.; Ng, Y.H.; Stubbs, L.P.; Torsæter, O. High Salinity and High Temperature Stable Colloidal Silica Nanoparticles with Wettability Alteration Ability for EOR Applications. *Nanomaterials* **2021**, *11*, 707. [CrossRef] [PubMed]
3. Alcorn, Z.P.; Føyen, T.; Gauteplass, J.; Benali, B.; Soyke, A.; Fernø, M. Pore- and Core-Scale Insights of Nanoparticle-Stabilized Foam for CO2-Enhanced Oil Recovery. *Nanomaterials* **2020**, *10*, 1917. [CrossRef]
4. Davarpanah, A. Parametric Study of Polymer-Nanoparticles-Assisted Injectivity Performance for Axisymmetric Two-Phase Flow in EOR Processes. *Nanomaterials* **2020**, *10*, 1818. [CrossRef] [PubMed]
5. Li, S.; Ng, Y.H.; Lau, H.C.; Torsæter, O.; Stubbs, L.P. Experimental Investigation of Stability of Silica Nanoparticles at Reservoir Conditions for Enhanced Oil-Recovery Applications. *Nanomaterials* **2020**, *10*, 1522. [CrossRef] [PubMed]
6. Rueda, E.; Akarri, S.; Torsæter, O.; Moreno, R.B.Z.L. Experimental Investigation of the Effect of Adding Nanoparticles to Polymer Flooding in Water-Wet Micromodels. *Nanomaterials* **2020**, *10*, 1489. [CrossRef] [PubMed]
7. Aadland, R.C.; Akarri, S.; Heggset, E.B.; Syverud, K.; Torsæter, O. A Core Flood and Microfluidics Investigation of Nanocellulose as a Chemical Additive to Water Flooding for EOR. *Nanomaterials* **2020**, *10*, 1296. [CrossRef] [PubMed]
8. Rezaei, A.; Abdollahi, H.; Derikvand, Z.; Hemma-Sarapardeh, A.; Mosavi, A.; Nabipour, N. Insights into the Effects of Pore Size Distribution on the Flowing Behavior of Carbonate Rocks: Linking a Nano-Based Enhanced Oil Recovery Method to Rock Typing. *Nanomaterials* **2020**, *10*, 972. [CrossRef] [PubMed]

Article

Experimental Investigation of Polymer-Coated Silica Nanoparticles for EOR under Harsh Reservoir Conditions of High Temperature and Salinity

Alberto Bila [1,2,*] and Ole Torsæter [3]

1. Department of Chemical Engineering, Eduardo Mondlane University (EMU), Av. Moç. km 1.5, Maputo CP. 257, Mozambique
2. Centre of Studies in Oil and Gas Engineering and Technology, Eduardo Mondlane University (EMU), Av. Moç. km 1.5, Maputo CP. 257, Mozambique
3. PoreLab Research Centre, Department of Geoscience and Petroleum, Norwegian University of Science and Technology (NTNU), S. P. Andersens veg 15a, 7031 Trondheim, Norway; ole.torsater@ntnu.no
* Correspondence: alberto.bila@uemep.uem.mz

Abstract: Laboratory experiments have shown higher oil recovery with nanoparticle (NPs) flooding. Accordingly, many studies have investigated the nanoparticle-aided sweep efficiency of the injection fluid. The change in wettability and the reduction of the interfacial tension (IFT) are the two most proposed enhanced oil recovery (EOR) mechanisms of nanoparticles. Nevertheless, gaps still exist in terms of understanding the interactions induced by NPs that pave way for the mobilization of oil. This work investigated four types of polymer-coated silica NPs for oil recovery under harsh reservoir conditions of high temperature (60 °C) and salinity (38,380 ppm). Flooding experiments were conducted on neutral-wet core plugs in tertiary recovery mode. Nanoparticles were diluted to 0.1 wt.% concentration with seawater. The nano-aided sweep efficiency was studied via IFT and imbibition tests, and by examining the displacement pressure behavior. Flooding tests indicated incremental oil recovery between 1.51 and 6.13% of the original oil in place (OOIP). The oil sweep efficiency was affected by the reduction in core's permeability induced by the aggregation/agglomeration of NPs in the pores. Different types of mechanisms, such as reduction in IFT, generation of in-situ emulsion, microscopic flow diversion and alteration of wettability, together, can explain the nano-EOR effect. However, it was found that the change in the rock wettability to more water-wet condition seemed to govern the sweeping efficiency. These experimental results are valuable addition to the data bank on the application of novel NPs injection in porous media and aid to understand the EOR mechanisms associated with the application of polymer-coated silica nanoparticles.

Keywords: polymer-coated nanoparticles; core flood; EOR; interfacial tension; wettability alteration; nanoparticle-stabilized emulsion and flow diversion

Citation: Bila, A.; Torsæter, O. Experimental Investigation of Polymer-Coated Silica Nanoparticles for EOR under Harsh Reservoir Conditions of High Temperature and Salinity. *Nanomaterials* **2021**, *11*, 765. https://doi.org/10.3390/nano11030765

Academic Editor: Fabien Grasset and Henrich Frielinghaus

Received: 29 January 2021
Accepted: 5 March 2021
Published: 18 March 2021

Publisher's Note: MDPI stays neutral with regard to jurisdictional claims in published maps and institutional affiliations.

1. Introduction

Water flooding is the most widely used secondary fluid injection process into an oil-bearing formation, after primary depletion, to improve oil recovery potential. Water is pumped from injection wells, sweeping the oil in the reservoir pores, to the production wells. In this course, water preferentially channels and flows through the high-permeability zones, leaving behind a significant amount of displaceable oil in low permeability-bypassed zones of the reservoir. The reservoir conformance problems manifest themselves or arise due to the contrasts in reservoir fluid properties, heterogeneity of reservoir permeability, fluid mobility contrast, etc. [1]. Moreover, during the productive life of an oilfield, these problems cause the oil to be easily trapped by capillary forces and/or bypassed by the oil recovery-drive fluid, resulting in excessive production of water and, therefore, resulting in poor sweep efficiency. The conformance problems, coupled with the scarcity of new oil field discoveries are the most pressing reasons for the emergence of new oil recovery technologies, aiming to (i) extract about 50% of the original oil in place (OOIP) that is left in the reservoir after primary and secondary recovery stages [2], (ii) increase oil production rates from existing fields, and (iii) fill the gap between energy supply and demand worldwide.

Nanotechnology has shown great potential to solve some of the above problems and increase profitability for the oil and gas companies. The building block of nanotechnology is the nanoparticle and it operates at the nanoscale. Nanoparticles (NPs) are defined as a collection of atoms bonded together with diameter size in the range of 1 to 100 nm [3]. Figure 1 illustrate a nanoparticle, it is composed of a *core*, the inner material, and the *shell*, the outer layer. The *core* determines the properties of a NP, whereas the *shell* provides a protective membrane and determines the solubility or binding affinity of the NPs with other materials [4]. For oil recovery applications, NP is designed to be wetted by both phases, thus be partly hydrophilic and partly hydrophobic [5]; together with the small size and the large surface area, NPs can have a profound displacement effect on the oil recovery-drive fluid.

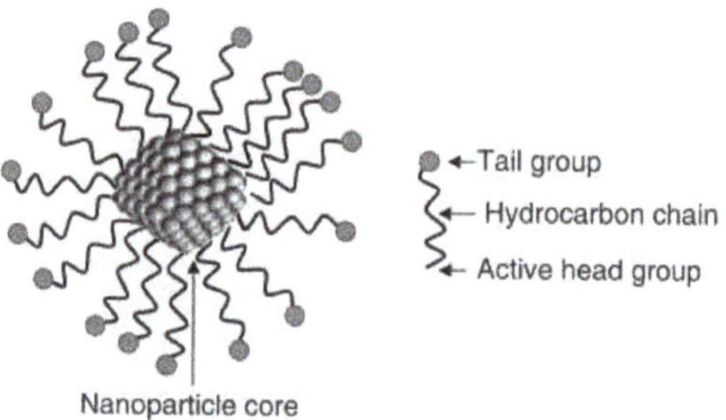

Figure 1. Schematic of a nanoparticle, showing the core and the shell [4].

The small particle size confers excellent mobility properties; hence, NPs can propagate deeply in the reservoir and increase oil recovery from thief zones and/or from bypassed zones with little retention. With small size, NPs have a large surface area and thus, a large contact area in the swept areas [6] and improved chemical reactivity properties. Therefore, NPs are suitable candidates for changing the reservoir rock and fluid properties and aid in the sweep efficiency of the injected fluid. In order to be successful for EOR applications, nanoparticles must (i) be stable in high salinity, high temperature, and high pressure reservoirs, (ii) propagate long distance between the injection and production wells with little retention, (iii) adsorb on the desired critical sites of the reservoir, such as the oil/water and fluid/rock interfaces, and (iv) prevent over-deposition on the pores [7].

One approach to achieving the above conditions and tailor the properties of NPs to improve their sweep efficiency, especially in harsh reservoir environments, is to covalently attach polymer molecules to the surface of the nanoparticles. The resulting novel polymer-coated NPs have received a wide interest in the oil and gas industry due to their improved solubility and stability, greater stabilization of emulsions and improved mobility through porous media [8,9]. Few studies have reported such characteristics for oil recovery. Rodriguez et al. [10] and Zhang et al. [11] reported that SiO_2 NPs coated with polymer molecules have a remarkable transport behaviour through reservoir pores of various permeability with little retention due to their reversible adsorption on the rock surface. Ponnapati et al. [12] experimentally found that polymer (*Poly-(oligo(ethylen oxide) monomethyl ether methacrylate)*-modified SiO_2 NPs could mobilise residual oil and yield 7.9% of the OOIP. Behzadi and Mohammadi [13] argues that polymer-coated silica NPs can modulate oil/water interfacial tension and change the wettability of an oil-wet glass micromodel to a more water-wet state, which can confer a greater EOR effect than unmodified silica nanoparticles. Choi et al. [14] reported that grafting polymer shell layers on the surface of silica NP can improve stability in harsh reservoir conditions. Their core flooding tests could achieve 74.1% of the OOIP with the modified NPs, which was quite comparable to plain water flood (68.9%) and unmodified silica NP. The authors associated the EOR effect to the NPs' ability to decrease the injection pressure; the authors argued that displacement pressure decrease is related to the formation of a wedge film between the oil and the rock surface. More recently, Bila et al. [15,16], Bila and Torsæter [17] carried out a series of flooding experiments with polymer-coated SiO_2 NPs in Berea sandstone core plugs. Their studies revealed an incremental recovery ranging from 2.6% to 14% of the OOIP. Then authors found that the displacement efficiency of polymer-coated SiO_2 NPs is better in water-wet cores than that achieved with induced neutral-wet core plugs.

In summary, the application of NPs, at laboratory scale, have shown an incremental recovery of oil ranging from 5 to 15% of the OOIP [3], the highest reach is 32% of the OOIP [18,19]. The most frequent range has been 5% of the OOIP [20]. Obviously, oil recovery by NPs is a complex phenomenon, partly because the reservoirs are unique and have different characteristics. On the other hand, the variability of experimental approaches in assessing the efficiency of nanoparticle recovery ends up with variable results and variable interpretation of the causes of oil displacement.

Following the encouraging results, studies have demonstrated marvellous efforts to understand the EOR mechanisms of nanofluid flooding. The change in reservoir wettability and the reduction of the interfacial tension are the two well accepted mechanisms of NPs [20,21]. Nanoparticles can alter the reservoir's wettability by (i) adsorbing on the reservoir rock to develop a new surface roughness [22,23], destabilizing oil films and desorbing it from the surface, and (ii) applying the structural disjoining pressure mechanism [24]. The adsorption of charged NPs can change the reservoir's wetting properties by forming hydrogen bonds with water molecules, there on attracting water molecules to the surface while lifting oil from it [18,25]. Further, the adsorption of NPs can reduce the interfacial energy between the rock surface and the oil, which disrupts whatever molecular attachment amongst the rock surface and the oil molecules attached to the surface [25]. Nanoparticles can also adsorb onto oil/water interface and decrease the interfacial tension between the two phases. For this, the NPs form a mono-layer that replaces the existing oil/water interface, acting as a mechanical barrier and bring the two phases together [26]. Moreover, depending on the hydrophobicity nature of NPs, they can irreversibly adsorb to the oil/water interface. There, the formed denser layer of NPs can protect oil droplets from flocculation and coalescence via steric mechanism to result in the stabilization of emulsion droplets by nanoparticles [9,27]. These emulsions (Pickering emulsions) can travel through the pores of the reservoirs with minimal retention and increase oil recovery [27].

Aside from the above mentioned mechanisms, NPs can block reservoir pore-throats larger than their size and increase oil recovery via log jamming mechanism [28,29]. Nanoparticles can increase the viscosity of EOR fluid and reduce the viscosity of the heavy oil, which can be reflected in a favourable mobility of the displacing and displaced phases, respectively, for oil recovery.

This work aims to expand our previous works remove the author's name only [15,16] and attempts to investigate the recovery mechanisms of polymer-coated silica NPs under harsh reservoir conditions of high temperature and salinity. The goal is to provide additional experimental results to the data bank on the nanoparticles for enhanced oil recovery projects, while paving the way to improve our understanding of the underlying EOR mechanisms associated with silica nanoparticles.

2. Experimental Materials and Methods

2.1. Nanoparticles and "Simulated" Seawater

Nanostructured products from Evonik Industries AG were used for oil recovery experiments in this work. The nanostructured materials were supplied as special research and development (R&D) laboratory products, hereafter simply referred to as nanoparticles. They were supplied to us in small containers as AERODISP®, which is AEROSIL® particles in liquid solution as shown in Figure 2. They are spherical and amorphous products mainly composed of silicon dioxide ($SiO_2 > 98.3\%$); additional components are aluminium oxide (Al_2O_3) and mixed oxides (MOX). The surface of NPs were coated with Poly (methacrylate) based molecules to improve dispersion stability and lower the hydrophilicity nature that characterizes typical silica materials to form polymer-coated/functionalized nanoparticles. Properties of the particles are given in Table 1. The size of the nanoparticles was measured using the dynamic light scattering technique.

Figure 2. Concentrated aqueous solutions of silica nanoparticles.

Table 1. Properties of the aqueous solutions of silica nanoparticles.

Sample	Basis	Modification	Conc., wt.%	Size (nm) [a]	Size (nm) [b]
NF-A	SiO_2 (sol-gel-cationic)	Polymer	38.6	107	63
NF-B	SiO_2 (sol-gel-anionic)	Polymer	26.0	32	38
NF-C	$SiO_2/Al_2O_3/MOX$	Polymer	21.6	218	155
NF-D	$SiO_2/Al_2O_3/MOX$	Polymer	25.5	145	135

[a] Average particle size in concentrated solutions (as received). [b] Average particle size measured with NPs diluted to 0.1 wt.% with simulated seawater.

The simulated seawater was prepared with ions typical of sea water (Na^+, Ca^{2+}, Mg^{2+}, SO_4^{2+}, K^+, etc.), by mixing salts with distilled water, following the homogenization process. The total of dissolved salts was 38,318 ppm. For oil recovery experiments, the concentrated solutions of nanoparticles were diluted to 0.1 wt.% concentration with simulated seawater. The resulting nanofluid solutions were stirred using a magnetic stirrer for at least 30 min to prevent aggregation/agglomeration of particles. The properties of water and nanofluids are given in Table 2, and were measured in our previous work [16]. The density and viscosity were measured at 60 °C with an Anton Paar density meter DMA^{TM} 4100 M series and Anton Paar Rheometer, respectively. The pH of the solutions was measured by a pH Meter (model pH 1000 L, phenomenal$^®$).

Table 2. Fluid properties measured at 60 °C.

Fluid	Density (g/cm^3)	Viscosity (cP)	pH
Seawater	1.008	0.53	7.97
Nanofluids	1.007–1.009	0.51–0.67	7.74–8.05

2.2. Non-Wetting Phase

Crude oil obtained from a field in the North Sea was used as the non-wetting phase, it had a viscosity of 6 cP and 33 °API gravity at at 60 °C. The crude oil SARA (Saturates, Aromatics, Resins, Asphaltenes) analysis is provided in Table 3. It is classified as a light oil with asphaltenes content of 0.18 wt.%. It was filtered twice through a 5 μm Millipore to remove suspended particles and preserve original composition. The density and viscosity were measured with the same instruments used in Section 2.1. Normal decane with density and viscosity of 0.73 g/cm^3 and 0.92 cP at 20 °C, respectively, was used in the wettability experiments.

Table 3. Properties of the crude oil.

Property	Value	Unit
Saturates	71.57	wt.%
Aromates	20.81	wt.%
Resins	7.44	wt.%
Asphaltenes	0.18	wt.%
Density at 60 °C	0.87	g/cm^3
Viscosity at 60 °C	6.0	cP
API gravity	33	deg

2.3. Preparation of Porous Medium

Eight core plugs were drilled from Berea sandstone block and prepared to have similar diameter (3.75 cm) and length (10 cm). The bulk mineral composition was measured using X-ray diffraction (XRD) on five core plugs. All samples were composed of 93.7 vol % quartz, 5 vol % of Microcline (Alkali feldspar) and 1.3 vol % Diopside. They are classified as homogeneous and strongly water-wet rocks. The cores were cleaned with methanol using Soxhlet extractor and dried at 60 °C for ≈3 days. Porosity was measured by imbibition method and permeability by Darcy's law using a constant head permeameter, flowing nitrogen gas through the core. The gas permeability obtained was then corrected for the Klinkenberg effect to reflect the liquid's permeability. The core properties are given in Table 4.

Table 4. Properties of the rock core plugs.

Core	Porosity, (%)	Permeability, mD	Pore Volume, mL	S_{wi} (%)
H1	17.61	332	19.25	24.66
H2	19.55	384	19.06	21.32
H3	17.40	361	19.22	16.24
H4	17.56	434	19.39	16.45
H5	16.70	460	19.83	17.81
H6	17.16	411	19.10	15.00
H7	17.61	425	19.25	14.00
H8	19.42	377	18.78	19.60

The cleaned and dried core plugs were saturated with reservoir simulated formation water using a vacuum pump with a chamber pressure set at 100 mbar for three hours. Then, the core plugs were soaked in the same water for at least 10 days to attain ionic equilibrium with the rock constituents.

Primary drainage was conducted by injecting crude oil into the cores to displace the water and set initial water saturation (S_{wi}). Next, the cores with S_{wi} were submerged in metallic containers filled with the same crude oil used to displace the water and soaked for seven months at 80 °C. The aim was to decrease the water wetness in the cores. The aging time and the high temperature can break the water films on the surface, facilitating the adsorption of polar components of crude oil. to change the rock wettability. The results of the aging process are presented in Section 3.2.4 and show that wettability of the cores has been changed from a strongly water-wet surface to a neutral-wet state.

2.4. Displacement Tests

Oil displacement tests were carried out at the reservoir temperature (60 °C) in aged core plugs to determine the EOR potential of silica based nanofluids. A schematic of the core flooding apparatus is shown in Figure 3. It is comprised of an injection pump, three tanks each containing crude oil, water and nanofluid. The reservoir tanks were assembled vertically inside an oven set at 60 °C. A check valve and back pressure regulator (set to 5 bars) were used to prevent any back flow of produced fluids and keep the system pressure constant during the experiments. The core was loaded in the Hassler core-holder and oriented horizontally under confining pressure held within 18–22 bars. The crude oil was injected at low rate (0.02 mL/min) to eliminate any air bubbles, until the temperature in the reservoir stabilized at 60 °C. To simulate oil production of a reservoir: (i) conventional water-flood was conducted at a constant reservoir flow rate, 0.2 mL/min, until there was no oil production for 1-pore volumes (PVs). Next, the flowrate was increased ten-fold (bump flood) for 1-PV in order to mitigate capillary end-effects, establish water flood residual oil saturation and ensure that any production would result from the injection of nanofluid. (ii) the injection was continued with nanofluids at the constant flow rate until there was no more oil production at core outlet for ≈3-PVs; likewise, the bump flood was applied for 1-PV. To study the behaviour of the injection fluid flow through the pores, two pressure sensors were connected at the entrance and exit of the core-holder; the differential pressure of the fluid across the core was recorded using a computer.

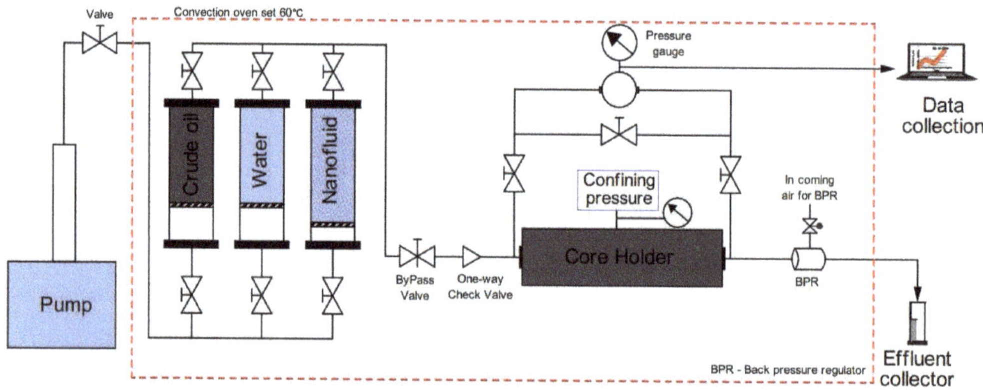

Figure 3. Schematic diagram of core flooding experiment.

2.5. Interfacial Tension Measurement

Interfacial tension (IFT) is the most important property that characterises the interface of two immiscible or partially miscible fluids that are in contact. Accordingly, it influences the extraction of crude oil from reservoir pore spaces. Interfacial tension can be measured by a variety of methods. A spinning drop method, SVT20N (Data Physics) video tensiometer, was used in this work. The crude oil was injected drop-wise into a capillary tube filled with water or nanofluid. Then, the apparatus was set a 60 °C and the tube rotated at a speed held within 6000–8000 rpm until the equilibrium was reached; then, the static IFT value was read off. The average refractive index of the continuous phases was 1.338.

2.6. Spontaneous Imbibition Tests

The spontaneous imbibition (SI) tests provide a qualitative measure of the ability of the wetting-phase to displace the non-wetting phase under static conditions [30]. Generally, the SI tests are performed in visual glass with graduated tube on top called Amott cell. An oil-saturated core plug (at S_{wi}) is place in the cell filled with water. Free-oil displacement by water is expected to take place over a period of time. The produced oil is collected on the top of graduated tube until equilibrium is reached. The dynamic imbibition is established as cumulative oil production versus time to interpret the change in wettability. In this work, the Amott water index (I_w) was also calculated to validate SI tests. At equilibrium, the volume of oil produced by SI is V_{o1}. The remaining oil in the core was forcibly displaced by injecting water; the produced oil is V_{o2}. The Amott water index is calculated as: $I_w = V_{o_1}/(V_{o_1} + V_{o_2})$. The closer to 1.0 the I_w is, the strongly water-wet the rock system is; smaller values of I_w indicate less water-wetting preference. It is worthy to mention that the core plugs were injected with decane before and after the nanofluid-EOR tests to set S_{wi}; then the imbibition tests were followed. This procedure aimed to study the wetting conditions that can be achieved after the interaction between the NPs and crude oil/water/rock system. The results are used to interpret oil displacement mechanisms of polymer-coated silica NPs.

3. Results and Discussion

3.1. Nanoparticle's Oil Recovery

Most oil reservoirs are characterised by high pressure, high temperature, high salinity and an uneven properties [8,31,32]. Thus, EOR technologies must be designed to operate nearly under identical conditions to provide realistic results. Presently, an attempt was made in this direction, carrying out the EOR experiments at the reservoir temperature of 60 °C, and dispersing the nanofluids in typical reservoir brine. The cores aged to less water-wet state. A bump flood was applied at the end of the low rate flood to ensure that any additional production would be a result of the nanofluid injection.

The oil recovery performance for the selected nanofluids is shown in Figure 4. Note that each nanofluid sample was tested twice. Figure 4a,b show the oil displacement profiles for the nanofluids with the smallest particle size (32 nm) and the largest particle size (218 nm), respectively. The oil recovery from the water flood (WF) is given by the blue curves, sequentially for (i) low rate (continued lines), and (ii) tenfold increase in flow rate for 1 PV (dashed lines). Then, the nanofluid was injected targeting water flood residual oil saturation. Likewise, the continued and dashed lines indicate the recovery of oil during the low rate and the high rate, respectively. The water flood conducted over the aged core samples exhibited an early water breakthrough, but accompanied with prolonged periods of co-production of water and oil. This displacement behavior conferred a greater oil recovery at the end of water flood, which is an agreement with previous studies [33,34]. The recovered oil, during the bump flood stage, seemed insignificant, except for the core H3 with 3.73% of the OOIP. This oil production can be attributed to the capillary end-effects and/or capillary stability. When capillary stability is not guaranteed during the low-rate displacement stage, the oil can snap off and easily become trapped in the larger pores as water saturation increases in the pores. These disconnected oil drops can be easily produced by increasing the flowrate. Experimental tests suggest that the capillary end-effects are less pronounced in intermediate rock systems [15,35], such as those used in this work.

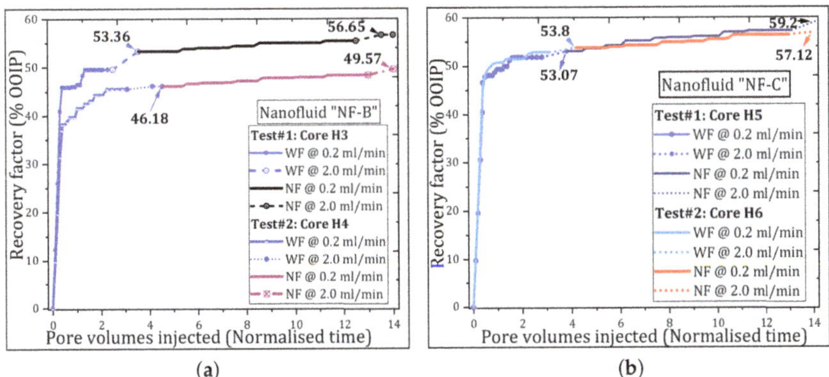

Figure 4. Effect of nanofluid injection on oil recovery during normal low rate and bump rate injections: (**a**) Sample NF-B: Test#1 the first oil production occurred ≈1.8 PVs and the RF ≈ 0.5% OOIP; Test#2 produced ≈ 0.5% OOIP at 1 PV. (**b**) Sample NF-C: Test #1 the oil production occurred at ≈2.2 PVs and the RF ≈ 2.3% OOIP; Test #2 the first production occurred at ≈1.3 PV and the RF ≈ 0.3% OOIP.

It is important to note that the current results indicate that on a laboratory scale, any EOR fluid must be performed after many pore volumes of water have been injected in the cores at low rate to achieve capillary stability and establish an adequate residual oil saturation. With this procedure, if the EOR fluid is successful in the laboratory, it is more likely to be successful in the field as well.

Nanofluid was injected to interact with the crude oil/rock/water system. The occurrence of oil at the exit of the core during the nanofluid injection stage was observed after 1 or 2 PVs (see Figure 4). This shows the time dependence of such interactions between the NPs and the crude oil/water/rock system towards improving the microscopic sweep efficiency of the oil-drive fluid. The interactions can be of particle adsorption type on the crude oil/water/rock interfaces. There on, the NPs can decrease the interfacial energy and change the wetting properties of the rock surface, and increase oil recovery. All results are summarised in Table 5. The average recoveries are shown in Figure 5 for comparison with the previous study. Total water flood oil recoveries varied in between 46.18% and 66.75% of the OOIP. This variation can be due to saturation profiles, pore structures, viscosity ratio and experimental errors. On average, 10-PVs of nanofluids were injected. This resulted in an incremental recovery ranging from 1.51 to 6.13% of the OOIP. The twin core plugs produced comparable recovery factors and the differences are in the margin of experimental error. The displacement efficiency (E_D) evaluated by Equation (1) shows the effectiveness of polymer-coated silica nanofluids in mobilising water flood residual oil at pore scale under harsh reservoir conditions.

$$E_D = \left[1 - \left(\frac{S_{or_2}}{S_{or_1}}\right)\right] \times 100\% \tag{1}$$

Here, the S_{or_1} and S_{or_2} is the residual oil saturation after water- and nanofluid-flooding, respectively.

Table 5. Oil recovery factors (water- and nanofluid-flooding), expressed as % of the OOIP, and residual oil saturation achieved at the end of core flooding in neutral-wet cores.

| Core | Fluid | Water Flood | | | | Nanofluid Flood | | | | | |
		RF1	RF2	RF	S_{or_1}	RF1	RF2	RF	S_{or_2}	RF_t	E_D (%)
H1	NF-A	59.45	2.90	62.35	28.37	2.76	2.07	4.83	24.47	67.18	13.74
H2		57.33	0.67	58.00	33.05	1.73	0.00	1.73	31.68	59.73	4.13
H3	NF-B	49.63	3.73	53.36	39.07	2.05	1.24	3.29	36.32	56.65	7.06
H4		45.56	0.62	46.18	44.97	2.16	1.23	3.39	42.14	49.57	6.31
H5	NF-C	51.84	1.23	53.07	38.57	4.29	1.84	6.13	33.63	59.20	12.81
H6		52.88	0.92	53.80	39.43	2.82	0.50	3.32	36.61	57.12	7.17
H7	NF-D	66.45	0.31	66.75	28.67	1.20	0.30	1.51	27.37	68.25	4.53
H8		53.64	0.80	54.44	37.27	3.05	0.93	3.98	34.08	58.41	8.57

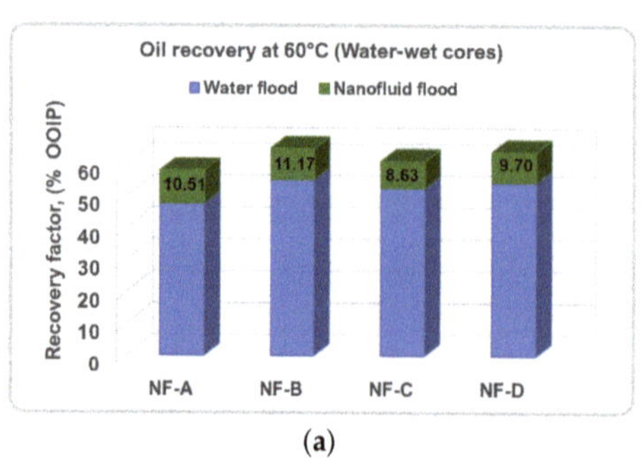

(a)

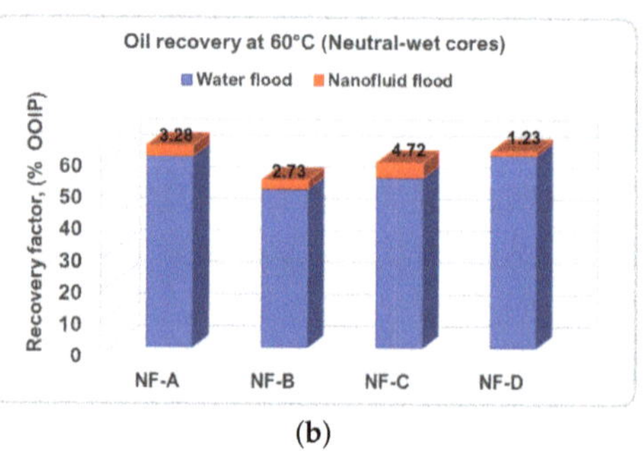

(b)

Figure 5. Comparison of average oil recoveries from various nanofluid floods: (**a**) Water-wet cores [16], and (**b**) Neutral-wet Berea core (present study).

The temperature seemed to have minor influence on oil recovery in neutral-wet cores, unlike in water-wet cores. Figure 5 illustrates the average oil recoveries for comparison purposes obtained in water-wet and neutral-wet Berea cores conducted under similar procedure and conditions. Figure 5 shows the nanofluid flooding results in water-wet systems and were extracted from Bila et al. [16], while the current results are given by Figure 5b and were calculated from Table 5. Clearly, the results indicate a superior oil recovery behaviour in water-wet Berea sandstone with the injection polymer-coated silica NPs. This may be associated with entrapping phenomenon in both rock systems during water flooding stage. In water-wet rocks, water fills the smaller pores and retains a considerable amount of the oil in the larger pores, which is amenable to EOR process. In contrast, neutral-wet cores tend to recover considerable amount of oil, despite the early water breakthrough. In addition, oil is trapped in the smallest pores and the residual oil saturation would resist mobilisation.

Aside from the temperature, the size of the nanoparticles seemed to influence the displacement efficiency; largely due to stability issues. Particularly, samples NF-C and NF-D with the largest particle size aggregated at core entrance during the EOR experiments. This resulted in physical filtration and the formation of nanoparticle "cake", shown in Figure 6. This was also observed in our previous work [16] but with water-wet Berea core plugs. However, the resultant layer of NPs appeared noticeably thicker in water-wet rocks, resulting in higher displacement pressures than in neutral-wet rock systems. This may indicate that polymer-coated NPs are more effective in improving the fluidity of the particles through neutral-wet Berea pores than in water-wet Berea pores.

Figure 6. Physical filtration and formation of nanoparticle "cake" at core inlet during nanofluid flooding.

3.2. Mechanisms behind EOR by Nanoparticles

3.2.1. Effect of Nanoparticles on Viscosity of Injection Water

Together with interfacial tension, the viscosity provides a qualitative measure of how, and to what extent, oil will flow through porous media [36]. The research have reported tremendous increase in the viscosity of the injection fluid in the presence of NPs, which can significantly improve oil sweep efficiency. The size and concentration of the NPs are the main parameters that affect the viscosity of the injection water [8,25]. An increase in the viscosity of injection fluid is reflected in the mobility of adjacent fluid molecules around the nanoparticles. For instance, highly concentrated nanofluid can give a uniform displacement front [12,25], which is favourable for increasing oil sweep efficiency.

In the present work, nanoparticles at 0.1 wt.% had negligible effect on the viscosity of injection seawater. The viscosity of the nanofluids varied from 0.51 to 0.67 cP compared to 0.53 cP of reference injection water. Hence, neither the concentration of polymers, as coating materials, nor the concentration of NPs was enough to generate a significant viscosity change in aqueous solution. The results lead to a similar conclusion to the previous study by Metin et al. [37], where low concentrations ($\leq$5 wt.%) of NPs have been shown not to affect aqueous viscosity. Therefore, we expect oil recovery to be largely influenced by factors other than viscosity.

3.2.2. Reduction in Interfacial Tension (IFT)

Normally, NPs are designed to be wetted by both phases, thus being partly hydrophilic and partly hydrophobic, to allow for a better reduction in IFT [5] and decrease capillary forces. At low capillary forces, the dynamic displacement lengthens the oil droplets within the pores; then the oil drops break into smaller ones to flow within the waterbed to the production wells.

Bila et al. [16] showed that IFT measurement with the Pendant drop method would not provide reliable results because, at high temperature, the polymer-coated SiO_2 NPs were attracted to the oil/water interface (see Figure 7a). From then on, making nanofluid solutions opaque before settlement due to gravity forces. Therefore, in this work, the IFT was obtained by analyzing the surface of the oil droplet immersed in a rotating capillary tube using a high-resolution camera and image analysis software. The use of a small volume of fluids in the capillary tube and the speed of rotation probably hindered the formation of large NP aggregates, allowing the measurement of IFT. At the equilibrium, IFT values obtained by the spinning drop method are given in Figure 7b as a function NP size.

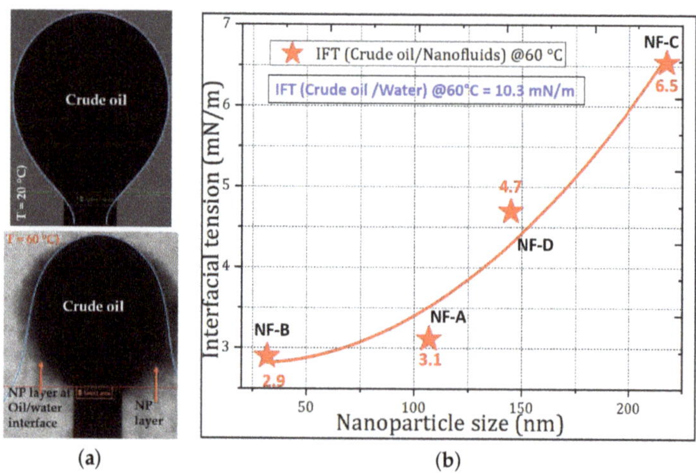

Figure 7. Dynamic measurement of the IFT between crude oil and nanofluids (0.1 wt.%): (**a**) Pendant drop method: Crude oil drop on top of a J-shape syringe needle at 20 °C (top) and at 60 °C (bottom). At 60 °C , the nanoparticles self-assembled onto the oil/water interface [16]. (**b**) Variation of the IFT measured with Spinning drop technique (at 60 °C) with nanoparticle size.

The IFT between crude oil and water, at 60 °C, was 10.3 mN/m [16]; in the presence of NPs, IFT decreased to 6.5 to 2.9 mN/m. Furthermore, Figure 7b shows that the reduction is dependent on the size of the nanoparticles. The smallest NPs tend to exhibit a high affinity capacity to adsorb at the interface, coming into contact with a large surface area; consequently, the NPs had the largest IFT reduction compared to the larger particles. For example, the NF-B sample with the smallest size (32 nm) reduced the IFT by 71.8% compared to the 36.9% reduction by the largest NF-C sample. This can be associated with the increase in repulsive forces between the smallest NPs, resulting in a greater disjoining pressure between the two-phase molecules [38]. In addition, the IFT reduction can be due to the surface-active polymer molecules surrounding the surface of the NPs that provide steric repulsive forces and stabilization of the NPs suspension. Comparing the results of the IFT data from previous studies at room temperature [15,17], for the same NPs, the greatest IFT reduction is achieved at elevated temperature. This indicates that the wetting properties and the binding energy of the NPs to the oil/water interface increase with temperature. This has been attributed to the intensification of Brownian motion, which increases the particle collision and reduces intermolecular interactions between oil/water [26]. However, the displacement of oil at elevated temperature still has a multiple effect on IFT, the formation of aggregates/agglomerates of NPs, the loss of NPs at the interface and the inhibition of NPs from performing their designated surface functions at the interfaces compete for poor performance of nanoparticles.

According to Sheng [39], significant oil recovery is achievable by reducing IFT to an ultra-low value of 10^{-3} mN/m. However, the reduction in IFT, even below the critical value, can still play a significant role in EOR; can alter the distribution of oil in the pores [40], can contribute to the generation and stabilization of emulsions by nanoparticles [8].

3.2.3. Formation of Emulsions Stabilized by Nanoparticles

A reduction in interfacial tension can create favorable conditions for the generation and stabilization of emulsion during the process of oil recovery by nanofluids. This premise was studied by increasing injection flow rate at the end of low rate injection. High flow rate can provide an extra energy necessary to break up oil phase and allow for NPs to adsorb to the fluids interface [41], thus generating emulsions. The produced core flooding effluent was then visualised using a high-resolution microscopic camera; the magnified images of the flooding effluent are presented in Figure 8. The images illustrate oil drops dispersed in aqueous solution in the presence of nanoparticles, indicating that residual oil was produced as oil-in-water emulsions.

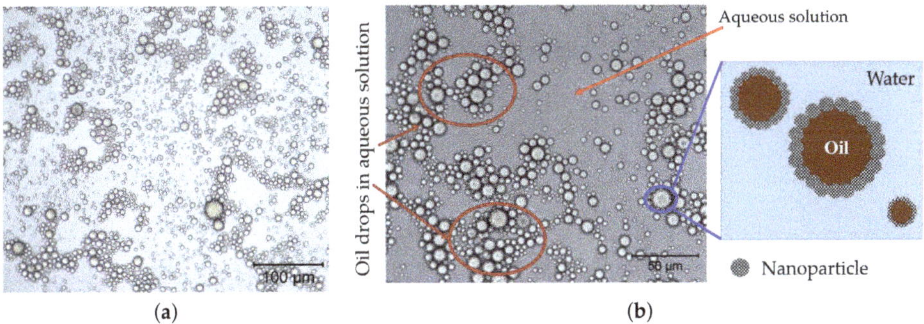

Figure 8. Visualisation of oil-in-water emulsion droplets from core flood effluent: (**a**) 20× magnification image, and (**b**) 50× magnification image.

In Figure 8, the adsorption of silica NPs seemed to create a rigid layer on the surface of the oil droplet, temporarily stabilising oil droplets against flocculation and coalescence via steric mechanism in the aqueous phase. The size distribution of the generated emulsions was not determined to compare with pore-throat sizes; however, Arab et al. [42] explain that oil drops created and stabilized by tiny NPs are small and possess a considerably lower viscosity than oil drops. Herewith, nanoparticle-stabilized emulsions, owing to their small size, can travel more easily through tiny reservoir rock pores with minimal retention, improved kinetic stability and negligible gravitational separation, resulting in increased oil recovery [8,23,26,43]. This profile mechanism (stabilisation of emulsions) probably occurred thanks to the polymer shells, as they temporarily prevented the coalescence of the oil drops. The phase separation of water and oil occurred approximately half an hour in the separator flask; the separated volumes were measured after two days.

The images showed in Figure 8 refers to emulsions generated in-situ by samples NF-A and NF-B; in contrast, samples NF-C and NF-D with the lager particle size generated weak emulsions. Among other factors, this shows that nanoparticles in the aggregate state are weak emulsifying agents. This has been attributed to either the kinetics of particles adsorption to the oil/water interface, being reduced by the presence of aggregated particles or networks, thus hindering the formation of emulsions [44].

3.2.4. Change in Rock Wettability and Surface Roughness

Nanoparticles target improving oil recovery by changing the surface roughness of the reservoir pores and thus altering its wettability to more water-wet state. The degree of water-wetness that can be achieved largely depend on how the NPs affect the crude oil/brine/rock properties [30]. Presently, we conducted spontaneous imbibition (SI) tests to evaluate such interactions induced by nanoparticles after nanofluid oil recovery experiments. In SI tests, the water will imbibe in the reservoir pores and displace the oil, if the reservoir is likely water-wet, as it has a positive capillary pressure [30]. The results of SI tests are presented in Figure 9. Prior to nano-EOR tests, the cores were aged 7 months in crude oil; post aged core wettability was evaluated by SI using one core plug. The lower curve, in Figure 9, shows the water imbibition performance in the aged core; there was no appreciable oil production before 10 days. Its Amott wettability index was −0.08, indicating neutral-wet condition. Additional cores aged under similar conditions were used for recovery tests, assuming that they retained the same neutral-wettability. After nanofluid core flood ceased production, the wettability alteration due to the adsorption of NPs was investigated. Figure 9 shows the cumulative oil production by water imbibition as a function of time. Depending on the particle size and particle retention in the pores, rock pore size distribution, etc., varied water imbibition profiles were achieved, but all showed an improved rate of water intake. That is, the water spontaneously invaded the rock pores and displaced the oil significantly, shortly after the placement of the core in the Amott cell. In contrast to the reference core where the oil production by water imbibition occurred from the tenth day, and reached stability 5 days later; thereafter, no oil production was observed. Part of the initial oil production of the aged core was the oil strongly attached to its surface after aging process. The jump in oil recovery, at 120 h, is because the cell was slightly shaken. In Figure 9, cores H2 and H4 were treated/injected with samples NF-A and NF-B, respectively. Its imbibition curves indicate superior oil recovery behavior by SI compared to other; this predicts the greater ability of the NPs to change oil-wet pores to water-wet state and increase oil recovery in a timely manner. It is worthy to mention that NF-A and NF-B had the smallest NP size and were stable throughout the duration of nanofluid flooding experiments. Therefore, with small size, NPs could contact most of the pore spaces to change its wettability.

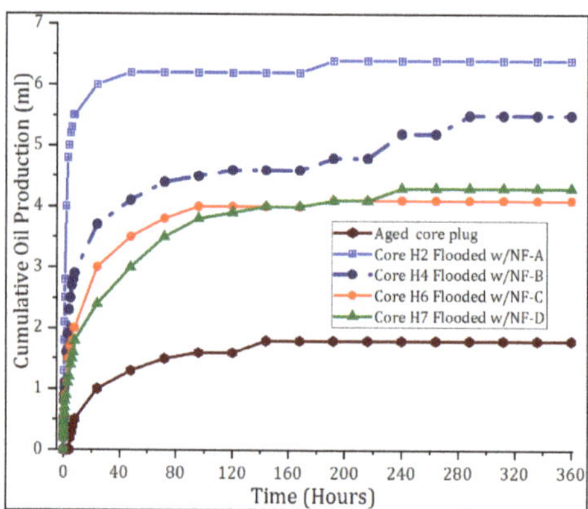

Figure 9. Spontaneous imbibition curves, before and after nanofluid core flooding.

In Figure 9, we see that greatest recovery of oil occurred within 40 h; after 40 h, an increase in oil recovery was observed. This indicates that the wettability was changed with time, in some parts of the pores, due to the nanoparticles. At equilibrium, the Amott water index (I_w) was determined and varied from 0.67 to 0.77. The rate of SI and the I_w, both indicate an alteration in core wettability to increasing water-wet condition. The capillary pressure that was negative in oil-wet pores was changed to positive values due to adsorption of NPs, which led to a stronger water imbibition in smallest pores. These results are in line with our previous study [15], and validate the ability of polymer-coated silica nanoparticles to alter the wettability of the rock to favor EOR process.

The mechanism behind the wettability alteration in oil-wet surfaces is still complex. However, it is probably associated with the adsorption of silica NPs to the crude oil/water/rock interfaces due to the attractive forces originating from dipole-dipole interactions [18,25] and structural disjoining pressure. These interactions can develop new surface roughness [23], decrease interfacial energy between the rock surface and the water [25], and destabilise oil films attached on the surface. Consequently, water-wet surfaces are created in the pores and greater capillary pressure ensures greater efficiency in oil recovery by nanoparticles.

3.2.5. Nanofluid Displacement Pressure

To better understand the mechanisms of oil recovery by nanofluids, dynamic curves of the pressure difference through the core plugs were recorded and analysed. The premise is that nanofluids can change the flow pattern or the pore volume available for flow; for this, the resulting differential pressure must be greater than the water flood pressure. Under these conditions, mobilization of residual oil can occur through so-called log-jamming effect.

Figure 10 shows the change in the flow pattern of the displacement pressure due to nanoparticles. While Figure 10a,b exhibit a slight change in water flood pressure, Figure 10c,d show a notable influence of NPs on the displacement pressure. The pair of the cores flooded with NF-B (Figure 10a,b) exhibited an injection pressure pattern similar to that of cores injected with the NF-A sample; the small size and dispersion stability of both samples (NF-A and NF-B) can be the reasons behind the negligible effect on the water flood displacement pressure. Furthermore, it suggests that both NF-A and NF-B samples can propagate through the pores and assist in oil recovery even in the smallest pores. The transport properties of nanoparticles were probably improved by the surface coating materials (polymer molecules), as they can eliminate electrostatic interactions between NPs and oil-wet surfaces [10,11]. The observed pressure profile mechanism during NP injection is consistent with the generation of in-situ emulsions. The propagation of NPs in smaller pores would release the oil from the surface to flow into the water bed and improve the relative permeability to the oil.

In contrast to samples NF-A and NF-B injection, there was a noticeable increase in water injection pressure when nanofluids with the largest particle size (NF-C and NF-D) were injected into the cores. In that case, the pressure gradually increased up to a point; there, it started to fluctuate (see Figure 10c,d). This displacement pressure pattern is a direct evidence of pore channel plugging. The NPs were partially transported through the pores, while others were retained, blocking the pores and forcing injection water to flow through through the bypassed pores. The diversion of water flow can create additional pressure in the adjacent pores; if the pressure is high enough to generate enough capillary numbers, the bypassed oil can be mobilized to the production wells and increase oil recovery. This displacement mechanism has been reported in the literature [28,29], and in most cases controls the oil displacement in water-wet cores [16,45,46].

Figure 6 shows the formation of NP "cake" as a result of aggregation/agglomeration of samples NF-C and NF-D at core inlet. The "nano-cake" is an evidence that the primary blockage occurred at the entrance of the core, followed by filtration of smaller NPs in response to the applied displacement pressure. Therefore, this phenomenon can be the main reason for the pressure and may not reflect oil recovery by the log-jamming effect.

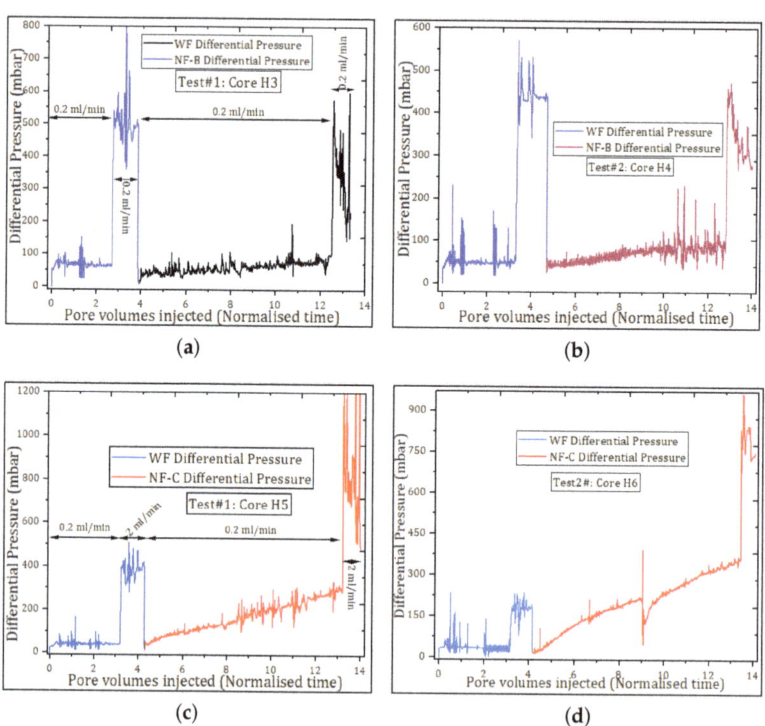

Figure 10. Pressure profiles for water flood (WF) and nanofluid flood oil recovery. (**a**) Test #1: Nanofluid "NF-B" injection showing little effect on water flood pressure; (**b**) In repeated Test #2, the pressure shows a slight increase compared to WF pressure; (**c**) Test #1, the pressure gradually increases with nanofluid "NF-C" and is higher than WF pressure; and (**d**) Similar behavior is observed in Test #2.

Inspecting the magnitude of the pressure increase due to NPs in Figure 10, it seems insufficient for the viscous forces to dominate the displacement process over the capillary forces, therefore, in this work, the flow diversion mechanism cannot fully explain the EOR effect due to the injection of nanoparticles.

4. Conclusions

This work presents the results of the influence of polymer-coated silica nanoparticles to enhance oil recovery at harsh reservoir conditions of high temperature and high salinity. This included an investigation of the effect of nanoparticle interactions on the crude oil/water/water interfaces, in order to understand the phenomenon of oil recovery. On the basis of the experimental results, the following conclusions can be drawn:

- Polymer-coated silica nanoparticles have shown a strong ability to increase oil recovery after water flooding. The increment recovery was up to 6% of OOIP;
- The nanoparticles can reduce the oil/water interfacial tension at a concentration as low as 0.1 wt.%. The smallest nanoparticle size were more efficient in reducing tension due to the large contact area;
- The flooding experiments indicated that oil was produced as oil-in-water emulsion droplets; these emulsion droplets were stabilized by the nanoparticles.

- The adsorption of nanoparticles in oil-wet pores can reverse the negative capillary pressure to positive values and change the wettability to water-wet condition;
- The size of nanoparticles and the formation of large aggregates within the pores were observed to increase displacement pressure, resulting in poor oil recovery efficiency;
- Different oil displacement mechanisms, such as reduced IFT, change in wettability, generation of in-situ emulsions and change log-jamming effect can explain the oil recovery phenomenon of polymer-coated silica NPs in intermediate reservoirs. However, the wettability alteration to a more water-wet seemed to govern the oil displacement process.

This paper reveals the potential application of polymer coated silica NPs for oil recovery. Future studies must be directed to improve the stability of the nanoparticles. In addition, studies should also focus on characterizing the charge on the rock surface, the effect of nanoparticles bonding with coating materials (i.e., surface activity and reactivity) to predict their interactions with fluids and rock system during the oil recovery process and probably to determine the contribution of the components involved in the oil recovery process.

Author Contributions: Conceptualization, A.B.; methodology, A.B.; investigation, A.B.; writing—original draft preparation, A.B.; writing—review and editing, A.B. and O.T.; supervision, O.T.; project administration, O.T. All authors have read and agreed to the published version of the manuscript.

Funding: This research was funded by the Centre of Studies in Oil and Gas Engineering and Technology, Eduardo Mondlane University, NUIT: 500003545, Maputo-Mozambique, and PoreLab, Centre of Excellence grant number 262644, Trondheim-Norway, and Centre of Studies in Oil and Gas Engineering and Technology, Eduardo Mondlane University (EMU)-Mozambique.

Data Availability Statement: Restrictions apply to the availability of Nanomaterials used in this study. The nanomaterials were obtained from Evonik Industries as research and development products and may be available on request from the company.

Acknowledgments: The authors would like to thank the Research Council of Norway for financial support through PoreLab, Centre of Excellence, project number 262644. Special thanks goes to Evonik Industries for providing the nanoparticles used in this work. Maximilian Cornelius and Ulrich Fischer from Evonik Industries are acknowledged for useful discussions. We thank Eng. Estevão Tivane for the design of the flooding set-up.

Conflicts of Interest: The authors declare no conflict of interest.

References

1. Sydansk, R.D.; Romero-Zeron, L. *Reservoir Conformance Improvement*; Society of Petroleum Engineers: Richardson, TX, USA, 2011.
2. Fletcher, A.; Davis, J. How EOR can be transformed by nanotechnology. In Proceedings of the SPE Improved Oil Recovery Symposium, Tulsa, OK, USA, 24–28 April 2010.
3. Bera, A.; Belhaj, H. Application of nanotechnology by means of nanoparticles and nanodispersions in oil recovery—A comprehensive review. *J. Nat. Gas Sci. Eng.* **2016**, *34*, 1284–1309. [CrossRef]
4. Das, S.K.; Choi, S.U.; Patel, H.E. Heat transfer in nanofluids—A review. *Heat Transf. Eng.* **2006**, *27*, 3–19. [CrossRef]
5. Peng, B.; Zhang, L.; Luo, J.; Wang, P.; Ding, B.; Zeng, M.; Cheng, Z. A review of nanomaterials for nanofluid enhanced oil recovery. *RSC Adv.* **2017**, *7*, 32246–32254. [CrossRef]
6. Sandeep, R.; Jain, S.; Agrawal, A. Application of Nanoparticles-Based Technologies in the Oil and Gas Industry. In *Nanotechnology for Energy and Environmental Engineering*; Springer: Berlin/Heidelberg, Germany, 2020; pp. 257–277.
7. Yu, H.; Kotsmar, C.; Yoon, K.Y.; Ingram, D.R.; Johnston, K.P.; Bryant, S.L.; Huh, C. Transport and retention of aqueous dispersions of paramagnetic nanoparticles in reservoir rocks. In Proceedings of the SPE Improved Oil Recovery Symposium, Tulsa, OK, USA, 24–28 April 2010.
8. ShamsiJazeyi, H.; Miller, C.A.; Wong, M.S.; Tour, J.M.; Verduzco, R. Polymer-coated nanoparticles for enhanced oil recovery. *J. Appl. Polym. Sci.* **2014**, *131*. [CrossRef]
9. Gbadamosi, A.O.; Junin, R.; Manan, M.A.; Yekeen, N.; Agi, A.; Oseh, J.O. Recent advances and prospects in polymeric nanofluids application for enhanced oil recovery. *J. Ind. Eng. Chem.* **2018**, *66*, 1–19. [CrossRef]
10. Rodriguez, E.; Roberts, M.; Yu, H.; Huh, C.; Bryant, S.L. Enhanced migration of surface-treated nanoparticles in sedimentary rocks. In Proceedings of the SPE Annual Technical Conference and Exhibition, New Orleans, LA, USA, 4–7 October 2009.

11. Zhang, T.; Murphy, M.J.; Yu, H.; Bagaria, H.G.; Yoon, K.Y.; Nielson, B.M.; Bielawski, C.W.; Johnston, K.P.; Huh, C.; Bryant, S.L. Investigation of nanoparticle adsorption during transport in porous media. *SPE J.* **2015**, *20*, 667–677. [CrossRef]

12. Ponnapati, R.; Karazincir, O.; Dao, E.; Ng, R.; Mohanty, K.; Krishnamoorti, R. Polymer-functionalized nanoparticles for improving waterflood sweep efficiency: Characterization and transport properties. *Ind. Eng. Chem. Res.* **2011**, *50*, 13030–13036. [CrossRef]

13. Behzadi, A.; Mohammadi, A. Environmentally responsive surface-modified silica nanoparticles for enhanced oil recovery. *J. Nanopart. Res.* **2016**, *18*, 266. [CrossRef]

14. Choi, S.K.; Son, H.A.; Kim, H.T.; Kim, J.W. Nanofluid Enhanced Oil Recovery Using Hydrophobically Associative Zwitterionic Polymer-Coated Silica Nanoparticles. *Energy Fuels* **2017**, *31*, 7777–7782. [CrossRef]

15. Bila, A.; Stensen, J.Å.; Torsæter, O. Experimental Investigation of Polymer-Coated Silica Nanoparticles for Enhanced Oil Recovery. *Nanomaterials* **2019**, *9*, 822. [CrossRef] [PubMed]

16. Bila, A.; Åge Stensen, J.; Torsæter, O. Polymer-functionalized silica nanoparticles for improving water flood sweep efficiency in Berea sandstones. *E3S Web Conf.* **2020**, *146*, 02001. [CrossRef]

17. Bila, A.; Torsæter, O. Enhancing oil recovery with hydrophilic polymer-coated silica nanoparticles. *Energies* **2020**, *13*, 5720. [CrossRef]

18. Roustaei, A.; Moghadasi, J.; Bagherzadeh, H.; Shahrabadi, A. An experimental investigation of polysilicon nanoparticles' recovery efficiencies through changes in interfacial tension and wettability alteration. In Proceedings of the SPE International Oilfield Nanotechnology Conference and Exhibition, Noordwijk, The Netherlands, 12–14 June 2012.

19. Roustaei, A.; Saffarzadeh, S.; Mohammadi, M. An evaluation of modified silica nanoparticles' efficiency in enhancing oil recovery of light and intermediate oil reservoirs. *Egypt. J. Pet.* **2013**, *22*, 427–433. [CrossRef]

20. Ding, H.; Zhang, N.; Zhang, Y.; Wei, M.; Bai, B. Experimental Data Analysis of Nanoparticles for Enhanced Oil Recovery. *Ind. Eng. Chem. Res.* **2019**, *58*, 12438–12450. [CrossRef]

21. Aziz, H.; Tunio, S.Q. Enhancing oil recovery using nanoparticles—A review. *Adv. Nat. Sci. Nanosci. Nanotechnol.* **2019**, *10*, 033001. [CrossRef]

22. Karimi, A.; Fakhroueian, Z.; Bahramian, A.; Pour Khiabani, N.; Darabad, J.B.; Azin, R.; Arya, S. Wettability Alteration in Carbonates using Zirconium Oxide Nanofluids: EOR Implications. *Energy Fuels* **2012**, *26*, 1028–1036. [CrossRef]

23. Dai, C.; Wang, X.; Li, Y.; Lv, W.; Zou, C.; Gao, M.; Zhao, M. Spontaneous Imbibition Investigation of Self-Dispersing Silica Nanofluids for Enhanced Oil Recovery in Low-Permeability Cores. *Energy Fuels* **2017**, *31*, 2663–2668. [CrossRef]

24. Wasan, D.; Nikolov, A. Spreading of Nanofluids on Solids. *Nature* **2003**, *423*, 156–159. [CrossRef]

25. Afolabi, R.O.; Yusuf, E.O. Nanotechnology and global energy demand: challenges and prospects for a paradigm shift in the oil and gas industry. *J. Pet. Explor. Prod. Technol.* **2019**, *9*, 1423–1441. [CrossRef]

26. Yakasai, F.; Jaafar, M.Z.; Bandyopadhyay, S.; Agi, A. Current Developments and Future Outlook in Nanofluid Flooding: A Comprehensive Review of Various Parameters Influencing Oil Recovery Mechanisms. *J. Ind. Eng. Chem.* **2020**, *93*, 138–162. [CrossRef]

27. Kumar, N.; Gaur, T.; Mandal, A. Characterization of SPN Pickering emulsions for application in enhanced oil recovery. *J. Ind. Eng. Chem.* **2017**, *54*, 304–315. [CrossRef]

28. Skauge, T.; Spildo, K.; Skauge, A. Nano-sized particles for EOR. In Proceedings of the SPE improved oil recovery symposium, Tulsa, OK, USA, 24–28 April 2010.

29. Kokubun, M.E.; Radu, F.A.; Keilegavlen, E.; Kumar, K.; Spildo, K. Transport of Polymer Particles in Oil–Water Flow in Porous Media: Enhancing Oil Recovery. *Transp. Porous Media* **2019**, *126*, 501–519. [CrossRef]

30. Mohammed, M.; Babadagli, T. Wettability alteration: A comprehensive review of materials/methods and testing the selected ones on heavy-oil containing oil-wet systems. *Adv. Colloid Interface Sci.* **2015**, *220*, 54–77. [CrossRef] [PubMed]

31. Miranda, C.R.; Lara, L.S.d.; Tonetto, B.C. Stability and mobility of functionalized silica nanoparticles for enhanced oil recovery applications. In Proceedings of the SPE International Oilfield Nanotechnology Conference and Exhibition, Noordwijk, The Netherlands, 12–14 June 2012. [CrossRef]

32. Khalil, M.; Jan, B.M.; Tong, C.W.; Berawi, M.A. Advanced nanomaterials in oil and gas industry: Design, application and challenges. *Appl. Energy* **2017**, *191*, 287–310. [CrossRef]

33. Anderson, W.G. Wettability Literature Survey Part 2: Wettability Measurement. *J. Pet. Technol.* **1986**, *38*, 1246–1262. [CrossRef]

34. Ziauddin, M.; Montaron, B.; Hussain, H.; Habashy, T.; Seleznev, N.; Signer, C.; Abdallah, W. Fundamentals of wettability. *Schlumberger Oilfield Rev.* **2007**, *19*, 40–67.

35. Eide, Ø.; Fernø, M.; Nybø, S.; Graue, A. Waterflood Stability in SCAL Analysis. In Proceedings of the International Symposium of the Society of Core Analysts, Avignon, France, 8–11 September 2014.

36. Bila, A.L. Experimental Investigation of Surface-Functionalised Silica Nanoparticles for Enhanced Oil Recovery. Ph.D. Thesis, Norwegian University of Science and Technology (NTNU), Trondheim, Norway, 2020.

37. Metin, C.; Bonnecaze, R.; Nguyen, Q. The viscosity of silica nanoparticle dispersions in permeable media. *SPE Reserv. Eval. Eng.* **2013**, *16*, 327–332. [CrossRef]

38. Behera, U.S.; Sangwai, J.S. Interaction of Nanoparticles with Reservoir Fluids and Rocks for Enhanced Oil Recovery. In *Nanotechnology for Energy and Environmental Engineering*; Springer: Berlin/Heidelberg, Germany, 2020; pp. 299–328.

39. Sheng, J.J. Status of surfactant EOR technology. *Petroleum* **2015**, *1*, 97–105. [CrossRef]

40. Chatzis, I.; Kuntamukkula, M.; Morrow, N. Effect of capillary number on the microstructure of residual oil in strongly water-wet sandstones. *SPE Reserv. Eng.* **1988**, *3*, 902–912. [CrossRef]
41. Kim, I.; Worthen, A.J.; Johnston, K.P.; DiCarlo, D.A.; Huh, C. Size-dependent properties of silica nanoparticles for Pickering stabilization of emulsions and foams. *J. Nanoparticle Res.* **2016**, *18*, 82. [CrossRef]
42. Arab, D.; Kantzas, A.; Bryant, S.L. Nanoparticle stabilized oil in water emulsions: A critical review. *J. Pet. Sci. Eng.* **2018**, *163*, 217–242. [CrossRef]
43. Suleimanov, B.A.; Ismailov, F.; Veliyev, E. Nanofluid for enhanced oil recovery. *J. Pet. Sci. Eng.* **2011**, *78*, 431–437. [CrossRef]
44. Binks, B.P.; Whitby, C.P. Nanoparticle silica-stabilised oil-in-water emulsions: improving emulsion stability. *Colloids Surfaces Physicochem. Eng. Asp.* **2005**, *253*, 105–115. [CrossRef]
45. Hu, Z.; Azmi, S.M.; Raza, G.; Glover, P.W.J.; Wen, D. Nanoparticle-Assisted Water-Flooding in Berea Sandstones. *Energy Fuels* **2016**, *30*, 2791–2804. [CrossRef]
46. Aurand, K.; Dahle, S.; Torsæter, O. Comparison of oil recovery for six nanofluids in Berea sandstone cores. In Proceedings of the International Symposium of the Society of Core Analysts, Avignon, France, 8–11 September 2014; pp. 1–12.

nanomaterials

Article

High Salinity and High Temperature Stable Colloidal Silica Nanoparticles with Wettability Alteration Ability for EOR Applications

Nanji J. Hadia [1,*], Yeap Hung Ng [1,*], Ludger Paul Stubbs [1] and Ole Torsæter [2]

[1] Institute of Chemical and Engineering Sciences, Agency for Science, Technology, and Research (A*STAR), 1 Pesek Road, Jurong Island, Singapore 627833, Singapore; ludger_paul@ices.a-star.edu.sg

[2] Department of Geoscience and Petroleum, Norwegian University of Science and Technology (NTNU), 7031 Trondheim, Norway; ole.torsater@ntnu.no

* Correspondence: hadianj@ices.a-star.edu.sg (N.J.H.); ng_yeap_hung@ices.a-star.edu.sg (Y.H.N.)

Abstract: The stability of nanoparticles at reservoir conditions is a key for a successful application of nanofluids for any oilfield operations, e.g., enhanced oil recovery (EOR). It has, however, remained a challenge to stabilize nanoparticles under high salinity and high temperature conditions for longer duration (at least months). In this work, we report surface modification of commercial silica nanoparticles by combination of zwitterionic and hydrophilic silanes to improve its stability under high salinity and high temperature conditions. To evaluate thermal stability, static and accelerated stability analyses methods were employed to predict the long-term thermal stability of the nanoparticles in pH range of 4–7. The contact angle measurements were performed on aged sandstone and carbonate rock surfaces to evaluate the ability of the nanoparticles to alter the wettability of the rock surfaces. The results of static stability analysis showed excellent thermal stability in 3.5% NaCl brine and synthetic seawater (SSW) at 60 °C for 1 month. The accelerated stability analysis predicted that the modified nanoparticles could remain stable for at least 6 months. The results of contact angle measurements on neutral-wet Berea, Bentheimer, and Austin Chalk showed that the modified nanoparticles were able to adsorb on these rock surfaces and altered wettability to water-wet. A larger change in contact angle for carbonate surface than in sandstone surface showed that these particles could be more effective in carbonate reservoirs or reservoirs with high carbonate content and help improve oil recovery.

Keywords: enhanced oil recovery; nanotechnology for EOR; nanoparticles stability; reservoir condition

---

check for updates

Citation: Hadia, N.J.; Ng, Y.H.; Stubbs, L.P.; Torsæter, O. High Salinity and High Temperature Stable Colloidal Silica Nanoparticles with Wettability Alteration Ability for EOR Applications. *Nanomaterials* **2021**, *11*, 707. https://doi.org/10.3390/nano11030707

Academic Editor: Alberto Villa

Received: 31 January 2021
Accepted: 3 March 2021
Published: 11 March 2021

Publisher's Note: MDPI stays neutral with regard to jurisdictional claims in published maps and institutional affiliations.

1. Introduction

Stages of oil recovery from petroleum reservoirs typically include primary, secondary, and tertiary recoveries processes. After primary (due to natural pressure energy of the reservoir) and secondary recovery (typically water or gas injection to maintain reservoir pressure) processes become uneconomical, tertiary recovery (also known as enhanced oil recovery (EOR)) methods are employed to improve oil recovery. In EOR, properties of reservoir rock and fluids and their interactions are altered to favorable conditions to mobilize residual oil trapped in oil reservoirs [1]. Two common EOR processes include thermal EOR and chemical EOR methods. In thermal EOR, heat energy is applied to the reservoir, usually by hot water or steam injection, to reduce the viscosity of oil and thus increase its mobility [2]. In chemical EOR methods, chemicals such as alkalis, polymers, and surfactants are injected into reservoirs to create favorable crude oil/brine/rock interactions to increase displacement and sweep efficiencies. Polymers such as hydrolyzed polyacrylamide (HPAM) and xanthan gum are commonly used to increase the viscosity of injection water and thus improve oil/water mobility ratio for effective displacement of oil by water [3]. Surfactants are primarily used to lower the interfacial tension (IFT) between oil and water and thus decrease capillary forces to mobilize trapped oil at pore throats [4].

These chemicals, however, tend to degrade and therefore loose its effectiveness under high temperature and high salinity conditions usually present in oil reservoirs. Moreover, a significant portion of the injected chemicals adsorb on the rock surfaces which leads to loss of these chemicals and eventually increases the project costs. These specially designed formulations of polymers and surfactants for harsh reservoir conditions are costly and make chemical EOR projects economically less viable. One of the ways to reduce the cost associated with chemical EOR is to replace or combine expensive with cheap and yet effective materials or chemicals. In this regard, nanotechnology may prove to be promising due to rapid progress and advancements in novel nanomaterials for various industries ranging from biomedical to personal care to oil and gas.

In recent years, nanotechnology is gaining increasing attention in oil and gas industries. Many research articles have provided a comprehensive overview on potential applications of nanotechnology in this area [5–9]. For EOR applications, at laboratory scale, nanoparticles were studied as performance improvers for polymer [10,11] and surfactant [12,13] flooding processes, wettability modifiers [14–18], emulsion stabilizers [19–22], etc. Among various types of nanoparticles, silica nanoparticles (SNP) are extensively studied for EOR applications. For a successful EOR application, a long-term colloidal stability of nanoparticles in harsh reservoir conditions is very crucial. The nanoparticle dispersions tend to lose colloidal stability under high salinity and high temperature conditions due to van der Waals forces and formation of aggregates over time. The stability of nanodispersions can be improved by electrostatic or steric stabilization mechanisms [23] to make them suitable for EOR applications. Recently, (3-glycidyloxypropyl) trimethoxysilane (GLYMO), a hydrophilic silane, has been proposed by many researchers to stabilize nanodispersions in hostile environments [24–27]. Worthen et al. [24] compared silica nanoparticles of 7–20 nm size stabilized by three types of nonionic ligands namely, GLYMO, polyethylene glycol (PEG), and zwitterionic sulfobetaine (SB) in seawater and American Petroleum Institute (API) brine. Their results showed that GLYMO and SB ligands were able to stabilize nanodispersions up to 80 °C for over 30 days in pH 3.5 API brine. The stability at pH > 3.5 is not reported. GLYMO also provided colloidal stability for 3 days even at 120 °C. Jang et al. [25] achieved colloidal stability of silica nanofluid up to salinity of 20% at 90 °C by modifying silica nanoparticles with GLYMO. Their wettability tests on oil-wet carbonate rocks showed effective wettability alteration to neutral- and water-wet by modified silica nanofluids. Griffith and Daigle [26,27] have shown that GLYMO-modified silica nanoparticles helped stabilize oil-in-water Pickering emulsions.

Wettability of reservoir rocks is a crucial parameter that controls oil production and affects the oil recovery during waterflooding. Oil recovery from oil-wet reservoirs, e.g., carbonates, is generally lower than intermediate- or water-wet reservoirs [17,28]. In water-wet reservoirs, at a given oil saturation, the oil occupies larger pores and water relative permeability is lower, which results in higher oil recovery. In oil-wet reservoirs, the water breaks through early and makes waterflooding uneconomical due to increased water-cut after breakthrough. Wettability also affects the oil recovery from heterogeneous reservoirs with zones of high and low permeability. In the case of water-wet rocks, water easily imbibes into low permeability zones, thus less bypassing occurs and oil recovery increases. In oil-wet rocks, water cannot spontaneously imbibe into low permeability zones due to high capillary forces, thus bypassing low permeability zones and hence lower oil recovery. To improve oil recovery from such oil-wet reservoirs, it is necessary to alter the wettability of rock surfaces to intermediate- or water-wet. Surfactants are traditionally used as wettability modifying agents for oil-wet reservoirs. Different types of surfactants have been studied to alter the wettability of oil-wet reservoirs [29–32]. During surfactant flooding, the surfactant molecules adsorb on the rock surface by various mechanisms and render it water-wet. The mechanism of adsorption depends on the type of surfactant, e.g., anionic, cationic, or non-ionic [33].

Recently, nanoparticles have been extensively studied for their potential to alter the wettability of rock surfaces. Xu et al. [14] formulated an ultra-low interfacial tension

(IFT) nanofluid consisting of surfactant and silica nanoparticles for very low permeability (0.2–0.3 mD) reservoirs. Their wettability experiments demonstrated that the silica nanoparticles could effectively alter the wettability of the rock, making it become more water-wet with increasing silica nanoparticle concentration. Bayat et al. [15] used aluminum oxide (Al_2O_3), titanium dioxide (TiO_2), and silicon dioxide (SiO_2) nanoparticles to study the adsorption and the resulting wettability alteration of limestone surface at different temperatures. The results showed wettability change of intermediate-wet limestone to water-wet due to adsorption of nanoparticles on the surface. Maghzi et al. [16] performed experiments to study the effect of adding silica nanoparticles to a polymer solution on heavy oil recovery in a five-spot glass micromodel. Based on oil recovery and contact angle measurements, they concluded that adding silica nanoparticles to the polymer solution changed the glass surface from oil-wet to water-wet and resulted in 10% higher oil recovery than by polymer flooding only. Al-Anssari et al. [17] also found that silica nanoparticles have the ability to alter the wettability of calcite surfaces from oil-wet to strongly water-wet by irreversible adsorption. Recently, Kanj et al. [18] developed a carbon nanofluid system for EOR application in high temperature and high salinity carbonate reservoirs. Their contact angle experiments showed effective wettability alteration of carbonate rock surfaces from oil-wet to water-wet.

In this study, we report functionalization and surface modification of commercial colloidal silica nanoparticles for high salinity and high temperature conditions. We compared the long-term stability of GLYMO-, SBS-, and GLYMO-SBS-modified nanodispersions at high temperature by static and accelerated stability analyses methods. The accelerated stability analysis simulated long-term stability in shorter period by subjecting the samples to centrifugal force and therefore predict the stability duration. Finally, contact angle measurements were performed in high salinity conditions on neutral-wet sandstone and carbonate rock surfaces to wettability alteration capability of the nanoparticles and therefore its suitability for EOR applications. The results showed that the GLYMO-SBS modified nanodispersion exhibited excellent stability in high salinity brines at 60 °C for at least 6 months and altered the wettability of neutral-wet sandstone and carbonate rocks to water-wet.

2. Experimental

2.1. Materials

Levasil OF50, an aqueous dispersion of colloidal silica (solid content 15 wt%, specific surface area 500 m^2/g, pH 10) was supplied by Nouryon Asia Pte Ltd., Singapore. The particles were used as received for further modification. The average DLS (dynamic light scattering) size of the particles was about 22 nm. Moreover, [3-(*N,N*-Dimethylamino)propyl]trimethoxysilane (>96.0%(GC), TCI), 1,3-propanesultone (≥99%, New York, NY, USA, Sigma-Aldrich), acetone (99.8%, Extra Dry, Acros Organics, Geel, Belgium), acetic acid (glacial, ≥99.8%, Sigma-Aldrich, New York, NY, USA), and (3-glycidyloxypropyl)trimethoxysilane (GLYMO, ≥98%, Sigma-Aldrich, New York, NY, USA) were used as received.

Two types of brines, synthetic seawater (SSW) and 3.5% NaCl, were prepared by dissolving different salts in ultrapure water followed by vacuum filtration using 0.22 μ filter. The composition of SSW is given in Table 1. Berea, Bentheimer, and Austin Chalk rock samples (Kocurek Industries, Caldwell, TX, USA) of dimensions $20 \times 10 \times 5$ mm^3 were used for contact angle measurements. The mineral composition of the rock samples measured by X-ray diffraction (XRD) is shown in Table 2. A stock tank light crude oil (density: 0.852 g/cc; viscosity: 2.43 mPa·s at room temperature) was used for saturation and aging of rock samples and contact angle measurements. The saturates, aromatics, resins, and asphaltenes (SARA) composition of the crude oil was 38.5%, 56.1%, 5.3%, and 0.1%, respectively.

Table 1. Composition of synthetic seawater.

Salts	Concentration (g/L)	Salts	Concentration (g/L)
NaCl	27.03	$MgCl_2 \cdot 6H_2O$	11.23
$CaCl_2 \cdot 2H_2O$	1.76	Na_2SO_4	4.81

Table 2. Mineral composition of rock samples by XRD.

Mineral, %	Berea	Bentheimer [34]	Austin Chalk
Quartz	90.87	99.0	–
Albite	0.89	–	–
Sanidine	0.46	–	–
Muscovite	3.75	–	–
Kaolinite	1.92	0.7	–
Rutile	–	0.3	–
Clinochlore	0.97	–	–
Dolomite	1.13	–	–
Calcite	–	–	100

2.2. Methods

2.2.1. Synthesis of 3-(Dimethyl(3-(Trimethoxysilyl)Propyl)-Ammonio)Propane-1-Sulfonate (SBS)

The synthesis of zwitterionic silane, SBS, was adapted from Estephan et al. [35]. In a 100-mL Schlenk flask, 8 g of [3-(N,N-dimethylamino)propyl]trimethoxysilane was diluted with 40 mL of anhydrous acetone under argon. Then, 3.4 mL of 1,3-propane sultone was subsequently added using an argon-purged syringe. The reaction mixture was stirred vigorously for 6 h. White precipitate was then collected through filtration, and then washed with acetone for three times. The collected solid was immediately vacuum dried for 24 h and stored under argon before use.

2.2.2. Surface Functionalization of Silica Nanoparticles

For synthesis of surface-functionalized colloidal silica referred as OF50-M, as illustrated in Figure 1, 8 g of OF50 was added to a 20-mL vial. SBS (150 mg, 0.46 mmol) and GLYMO (150 μL, 0.68 mmol) were added to the colloidal dispersion under vigorous stirring. After the opaque liquid turned transparent, the mixture was stirred at 200 rpm for 24 h at room temperature. The modified silica nanodispersions were then diluted to 1 wt.% using 3.8% SSW or 3.5% NaCl, and the pH was adjusted to about 4 using glacial acetic acid.

Silica nanoparticles in the form of dry powder were prepared for chemical analysis. Pristine OF50 (OF50-UM) was diluted to ca. 5 wt.% with ultrapure water and centrifuged at 14,000 rpm (Allegra^{TM} 25R Centrifuge Beckman Coulter) for 2 h. The collected solid was washed twice with water, rinsed with methanol once, and dried under vacuum at room temperature for 2 days. OF50-M was dialyzed in DI water for 1 week using a dialysis membrane (Spectra/Pro$^{®}$7 MWCO 50 kD). The purified sample was subjected to similar centrifugation and washing procedure for attaining the pure solid. Similar to the modification protocol for OF50-M, two control samples, OF50-C1 and OF50-C2, were also prepared by functionalizing solely with SBS and GLYMO, respectively.

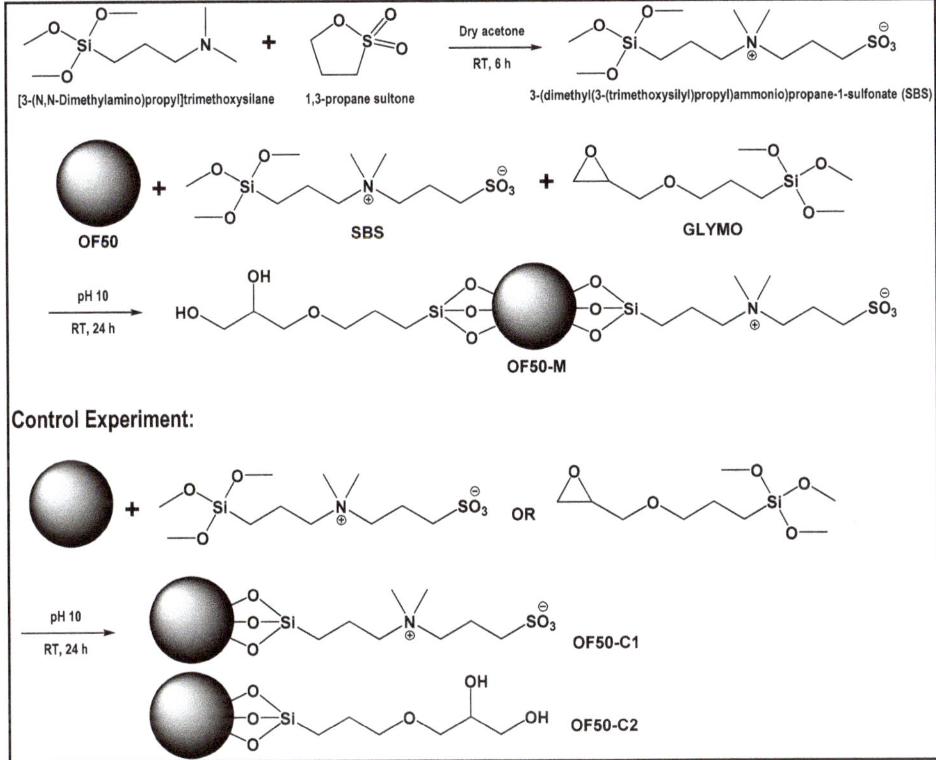

Figure 1. Synthesis of 3-(dimethyl(3-(trimethoxysilyl)propyl)-ammonio)propane-1-sulfonate (SBS) and surface modification of Levasil OF50 using SBS and (3-glycidyloxypropyl) trimethoxysilane (GLYMO).

2.2.3. Structural and Property Characterization

The Fourier-transform infrared (FTIR) spectra for pristine and functionalized silica were obtained from a PerkinElmer Frontier^{TM} FT-IR/NIR spectrometer with KBr as background. Particles size analysis of colloidal samples (concentration 1 wt.%) was done by dynamic light scattering (DLS) using Malvern Zetasizer Nano ZS. Particle morphology was observed using JEOL JSM7900F field emission scanning electron microscope (FE-SEM) at accelerating voltage of 1.5 kV under GBSH mode.

2.2.4. Stability Analysis of Nanoparticle Dispersions

The thermal and long-term stability tests of nanodispersions were performed (i) by visual observations, (ii) by turbidity scanning test using Turbiscan^{TM} instrument for real-time monitoring of transmission intensities over time, and (iii) using accelerated stability analysis. In turbidity scanning tests, the transmission profiles were recorded at every 3 h for 7 days. For this purpose, 1 wt.% OF50-M in SSW and 3.5% NaCl brines at pH 6 were subjected to turbidity scanning tests.

The accelerated stability analysis helps simulate long-term stability in shorter period of time by subjecting the samples to centrifugal force. A LUMiSizer instrument (LUM GmbH, Germany) was used for this purpose. In this study, the stability duration of 6 month was simulated in the runtime of about 31 h and transmission signals were recorded by a SEPView software [36] provided with the instrument. All accelerated stability analysis measurements were performed at 60 °C for 1 wt.% nanoparticles in SSW and 3.5% NaCl brine in the pH range of 4–7. For OF50 and OF50-C1, samples at pH 4 and 7 were used for accelerated stability analysis. All the samples were run in duplicates to counter-balance the weight on the rotor of the instrument. The size of nanoparticles was analyzed at room temperature for fresh samples as well as samples that were subjected to turbidity scan and accelerated stability analysis.

2.2.5. Contact Angle Measurements

In these experiments, the rock samples of dimensions $20 \times 20 \times 5$ mm^3 were first cut into two equal parts and washed thoroughly with ethanol, blow-dried, and placed in an oven at 70 °C overnight for complete drying. Figure 2 depicts the procedure followed for rock sample preparation for contact angle measurements. The dried chips were then saturated with the crude oil using vacuum saturation. The rock samples with 100% oil saturation were placed in an oven at 70 °C for aging to obtain representative wettability conditions. Berea and Bentheimer Sandstone rock samples were aged for 14 days, whereas Austin Chalk rock samples were aged for 2 days. At the end of aging period, all the rock samples were lightly rinsed with toluene to remove any free crude oil yet preserving the original wettability condition of the pore surfaces. The rock samples were then dried overnight in the oven at 70 °C. The dried rock chips were subsequently saturated with brines using vacuum saturation and used for contact angle measurements. To observe the wettability alteration by nanoparticles on the same rock sample, about half the rock sample was treated with nanofluids by immersing in the nanofluid for overnight and the other half was left untreated. It was thus possible to measure the contact angles on untreated and treated part of the rock chips without using duplicate rock chips.

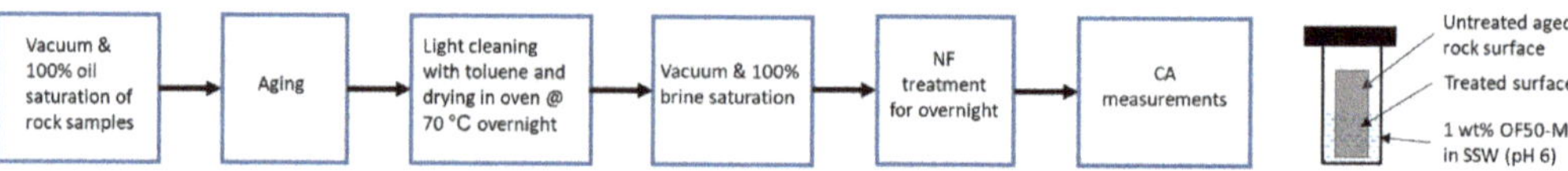

Figure 2. Procedure to prepare rock samples for contact angle measurements.

Contact angle measurements were performed at room temperature by captive bubble method using an optical contact angle meter (Model: OCA50, DataPhysics Instruments GmbH). For each measurement, a few hours were allowed for equilibration; all the measurement values reported herein were equilibrium values.

3. Results and Discussion

3.1. Nanoparticle Modification and Characterization

The FTIR spectra of pristine and functionalized silica particles are shown in Figure 3. For the pristine unmodified OF50-UM, strong absorption peaks at 469, 807, and 1100 cm^{-1} correspond to Si-O-Si bending (ν_b), the Si-O-Si symmetric (ν_s), and asymmetric (ν_{as}) stretching vibrations, respectively. For SBS and GLYMO functionalized OF50-M, small intensity peaks between 3000 and 2800 cm^{-1} are attributed to the stretching vibrations of the methylene groups from GLYMO, which confirms the successful surface chemical modification of OF50. For the identification of sulfobetaine functionality, the absorption peak at 1485 cm^{-1} corresponding to the C-N stretch from -N$^+$-(CH$_3$)$_2$ is used. The peaks arising from ca. 1220 cm^{-1} ν_{as}(SO$_3^-$) and 1040 cm^{-1} ν_s(SO$_3^-$) are not identified due to overlapping with strong Si-O-Si peak. Figure 4 shows SEM images of pristine OF50 and OF50-M nanoparticles. It can be observed that the size of the nanoparticles did not change significantly after the modification.

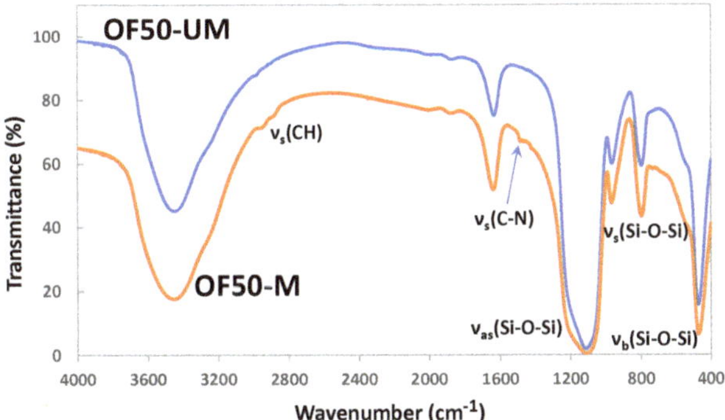

Figure 3. FTIR spectra of pristine OF50 (OF50-UM) and surface-modified/functionalized colloidal silica (OF50-M).

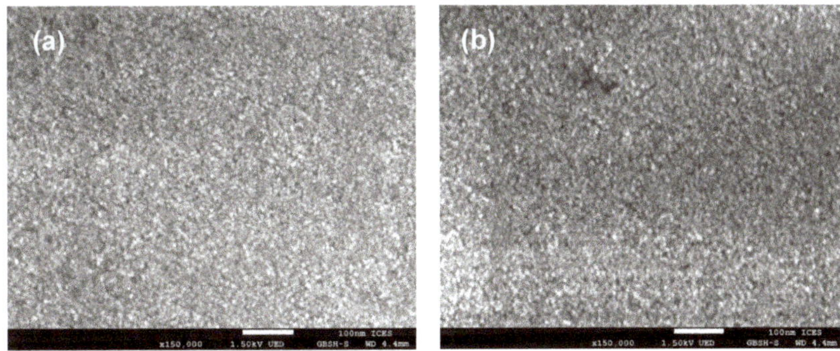

Figure 4. Scanning electron micrographs of (**a**) pristine OF50 and (**b**) OF50-M.

3.2. Nanoparticle Dispersion Stability

The thermal stability of unmodified, control, and modified nanoparticles dispersions was monitored at 60 °C for 30 days, and pictures were taken at regular intervals for visual observations. Figure 5 shows the photographs of the nanoparticle dispersions. It can be observed that OF50—UM nanoparticles precipitated out almost instantaneously in SSW, and after 3 h in 3.5% NaCl (Figure 4a). This shows that dispersions of the particles in their native form are not stable in high salinity conditions. The SBS-modified particles (OF50-C1) at pH 4 and 7 remained stable a little longer than unmodified nanoparticles but aggregated within 24 h, as can be observed in Figure 5b. Nanoparticles modified with GLYMO (OF50-C2) at pH 6 and 7 also aggregated within a day as can be observed in Figure 5d. These results show that only SBS or GLYMO modification is not enough to stabilize these nanodispersions in high salinity and high temperature conditions. On the other hand, dispersions of OF50-M nanoparticles showed excellent thermal stability in 3.5% NaCl as well as SSW in the pH range of 4–7 for 30 days without any sign of aggregation (Figure 5e). Modification of OF50 nanoparticles with combination of SBS and GLYMO helped to stabilize the dispersions for very long time. We anticipate that the optimum balance of surface coverage of hydroxyl and sulfobetaine functional groups played the pivotal role for ensuring the long-term saline stability of OF50-M nanodispersions. While the SBS-covered OF50-C1 was still susceptible to salt-induced destabilization due to presence of unmodified surface silanol group (-Si-OH), presence of grafted GLYMO was able to further enhance the solvation of the nanoparticles, which promoted the saline stability of dispersions at elevated temperature.

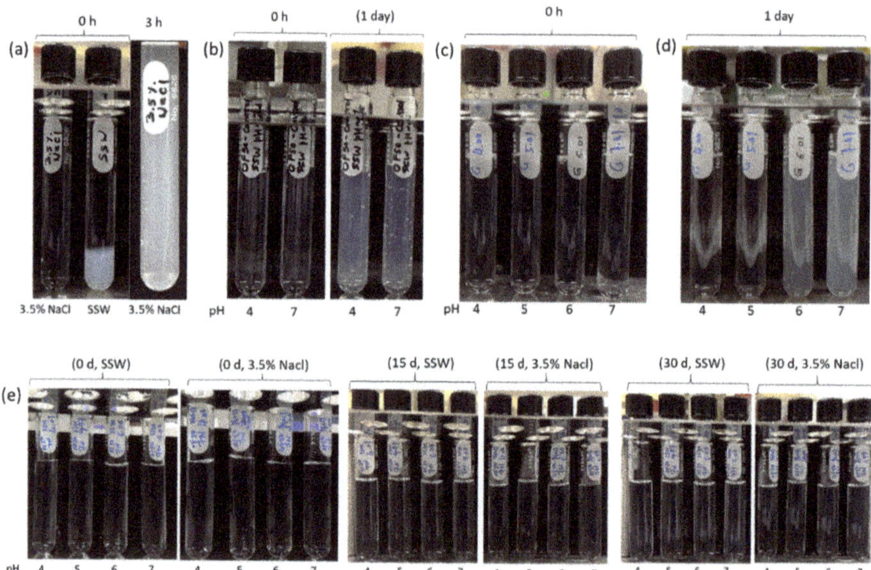

Figure 5. Thermal stability of 1 wt.% (**a**) OF50-UM in synthetic seawater (SSW) and 3.5% NaCl, (**b**) SBS modified particles (OF50-C1) in SSW at pH 4 and 7, (**c**) nanoparticles modified with GLYMO (OF50-C2) in SSW at pH 4–7 at 0 h, (**d**) OF50-C2 in SSW at pH 4–7 after 1 day, and (**e**) OF50-M in SSW and 3.5% NaCl brine at pH 4–7 for 30 days at 60 °C.

Results of a real-time turbidity monitoring of 1 wt.% OF50-M in SSW and 3.5% NaCl brine at pH 6 by Turbiscan^{TM} are shown in Figure 6. The results are presented in the form of transmission signals vs. height (0 mm represents bottom of the sample) of the sample for 7 days with each profile recorded every 3 h. No significant change in the transmission (ΔT) intensity was observed for 7 days across the height of the sample. This indicated that the particles were not aggregated and the dispersions remained stable over the duration of the test. The right side of the curves shows a shift of transmission profiles towards left (blue to red). This may be attributed to evaporation of the water in the sample vials as they were not completely filled with the nanofluid.

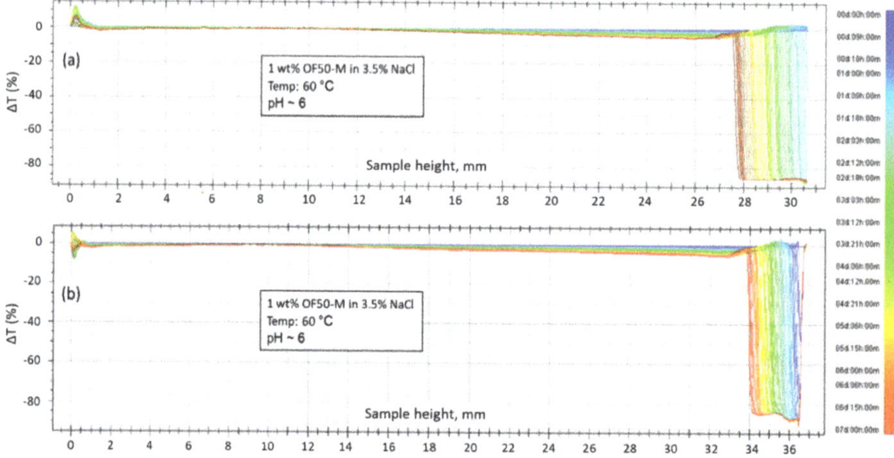

Figure 6. Transmission (ΔT) profiles obtained by Turbiscan^{TM} at 60 °C for 1 wt.% OF50-M nanoparticles in (**a**) SSW and (**b**) 3.5% NaCl brine for 1-week duration.

Results of accelerated stability analysis of OF50-UM, OF50-C1, and OF50-M nanofluids are shown in Figures 7–9 (with one repeat measurement for each nanofluid). OF50—C2 was not further tested due to insufficient colloidal stability. The time of measurement is indicated by color of the curves, from beginning (red) to end (green). The x-axis represents the height of the sample from top (130 mm) to bottom (105 mm). From Figure 7a–d, it can be observed that light transmission decreased with time from initial 80% to about 50% for nanofluid dispersions with 1 wt.% OF50-UM nanoparticles in 3.5% NaCl brine at pH 4 and 7. The decreased light transmission indicated aggregation of nanoparticles with time. For OF50-UM in 3.5% NaCl brine at pH 7, a sudden reduction in transmission from 80% to 5% occurred at about 124 mm (about 6 mm from bottom of the sample vial) as can be observed from Figure 7c,d. The formation of gel started after about 48 min of the test that is equivalent to simulated time of about 3 days. A significant reduction in transmission clearly showed a formation of gel and reduced the light transmission.

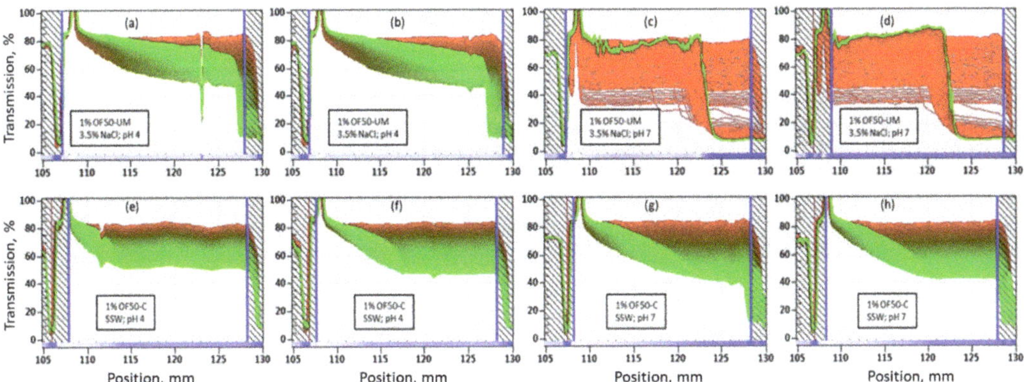

Figure 7. Accelerated stability analysis at 60 °C for 1 wt.% (**a**,**b**) OF50-UM in 3.5% NaCl brine at pH 4, (**c**,**d**) OF50-UM in 3.5% NaCl brine at pH 7, (**e**,**f**) OF50-C1 in SSW at pH 4, and (**g**,**h**) OF50-C1 in SSW at pH 7. The simulated stability duration was 6 months and runtime of experiment was 31 h.

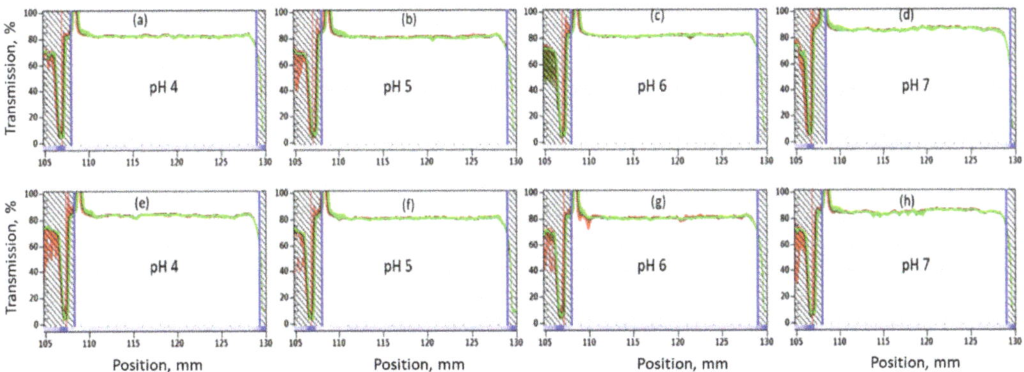

Figure 8. Accelerated stability analysis at 60 °C for 1 wt.% OF50-M in 3.5% NaCl brine at (**a**,**e**) pH 4, (**b**,**f**) pH 5, (**c**,**g**) pH 6, and (**d**,**h**) pH 7. The simulated stability duration was 6 months and runtime of experiment was 31 h. Samples were run in duplicate.

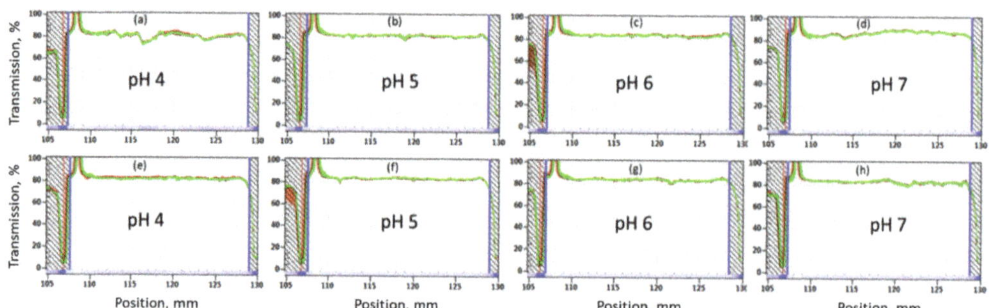

Figure 9. Accelerated stability analysis at 60 °C for 1 wt.% OF50-M in SSW at (**a,e**) pH 4, (**b,f**) pH 5, (**c,g**) pH 6, and (**d,h**) pH 7. The simulated stability duration was 6 months and runtime of experiment was 31 h. Samples were run in duplicate.

From Figure 7e–h, it can be observed that the nanofluids with 1% OF50-C1 also showed a decrease in % transmission from about 80% to 50% at pH 4 and to 40% at pH 7 during the runtime of 31 h. This also clearly shows that the surface modification with zwitterionic silane (SBS) only is not sufficient to stabilize the nanodispersions.

The transmission profiles of 1 wt.% OF50-M in 3.5% NaCl and SSW at different pH are shown in Figures 8 and 9. It can be observed that the transmission intensity did not vary significantly across the height of all the samples for experiment time of 31 h (equivalent to simulated stability time of 6 months). These results clearly indicate that 1wt% OF50-M nanoparticles remain stable for at least 6 months at 60 °C in both types of brines. This can be attributed to the synergy of SBS and GLYMO in stabilizing the nanoparticles against aggregation, as compared to nanoparticles with modification by SBS or GLYMO only.

The results of DLS measurements for OF50-M are shown in Table 3. It can be observed that the sizes of the nanoparticles were nearly the same in SSW and 3.5% NaCl brine. This indicates that the colloidal stability of modified nanoparticles was not affected by the presence of divalent ions in SSW. This combined effect of steric and electrosteric stabilizations imposed by GLYMO and SBS, respectively, is the key factor for enhanced stability of silica nanodispersions stability in high salinity brines. The particle size of samples that were subjected to turbidity scan (for 7 days) and accelerated stability analysis at 60 °C only marginally increased at pH 6 and 7, which further validates good stability of OF50-M particles in high salinity brine at high temperature.

Table 3. Dynamic light scattering (DLS) size measurements of 1 wt% OF50-M in synthetic seawater (SSW) and 3.5% NaCl brine when fresh and after turbidity scan and accelerated stability analysis measurements.

| | Particle Size, nm | | | | | |
| pH | Fresh Sample | | Treated Sample * | | | |
	SSW	3.5% NaCl	SSW-Turbi [+]	SSW-ASA [#]	3.5% NaCl-Turbi [+]	3.5% NaCl-ASA [#]
4	22.3 ± 0.1	21.6 ± 0.2	–	22.3 ± 0.1	–	21.6 ± 0.1
5	22.5 ± 0.2	21.7 ± 0.1	–	22.7 ± 0.1	–	21.9 ± 0.1
6	22.5 ± 0.2	21.7 ± 0.1	24.2 ± 0.1	24.0 ± 0.1	23.5 ± 0.2	22.9 ± 0.1
7	22.7 ± 0.1	21.8 ± 0.1	–	24.8 ± 0.1	–	22.9 ± 0.1

* Samples first used for turbidity scan and accelerated stability. [+] Sample after turbidity scan. [#] Samples after accelerated stability analysis.
– Samples were not subjected to turbidity scan.

The results of static and accelerated stability analysis and DLS measurements for fresh and treated samples (i.e., subjected static and accelerated stability tests) confirmed excellent stability of OF50-M nanodispersion in SSW and 3.5% NaCl brine at 60 °C. As mentioned earlier, the resulted stability can be attributed to the synergistic effects of GLYMO and SBS and optimum surface coverage of nanoparticles by GLYMO and SBS [35,37].

3.3. Wettability Alteration

Herein, the results of contact angle measurements are reported and discussed. As mentioned earlier, contact angles were measured on Berea Sandstone, Bentheimer Sandstone, and Austin Chalk rock samples. Figure 10 shows a comparison of contact angles measured on untreated and OF50-M treated sections of the rock chips. The values of contact angles are provided in Table 4. It should be noted that the contact angles are measured through the oil phase.

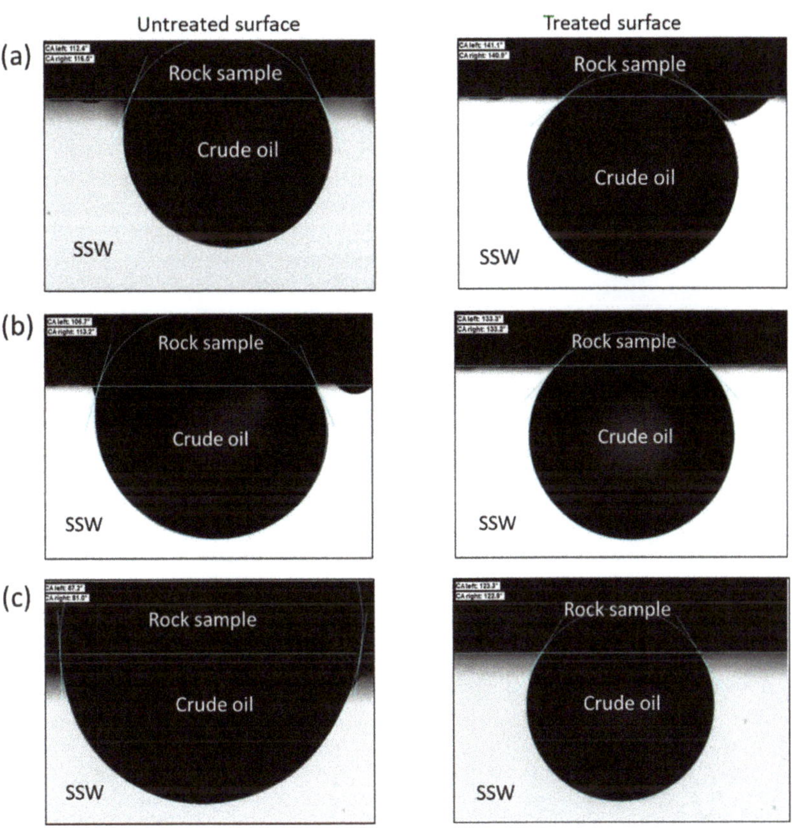

Figure 10. Contact angle measurements on untreated and nanofluid (1 wt.% OF50-M) treated surfaces of (**a**) Bentheimer Sandstone, (**b**) Berea Sandstone, and (**c**) Austin Chalk rock samples at room temperature.

Table 4. Contact angles measured on untreated and treated rock surfaces.

Rock Type	Contact Angle *	
	Untreated Surface	Nanofluid Treated Surface
Bentheimer	113°	141°
Berea	110°	131°
Austin Chalk	84°	123°

* mean contact angle.

It is well known that outcrop rocks such as the ones used in this study are naturally strongly water-wet. When aged in crude oil, the polar component from the crude oil adsorbs on the pore walls and change the water-wet rocks to neutral- or oil-wet depending on the crude oil composition and aging time. From Figure 10, it can be observed that Bentheimer and Berea Sandstone samples attained a weakly water-wet condition, whereas neutral-wet condition was observed for Austin Chalk sample after aging in crude oil. When part of these rock samples treated with 1 wt.% OF50-M, the wettability of the surfaces changed to water-wet for all the rock samples. This wettability alteration can be attributed to the adsorption of nanoparticles on the rock surfaces. Due to overall hydrophilic nature of the particles, the surfaces exhibit hydrophilic (water-wet) behavior after the treatment with nanofluid. The change in contact angles for Bentheimer, Berea, and Austin Chalk samples was found to be 28°, 21°, and 39°, respectively (Table 3). Sandstone pore surfaces are negatively charged, whereas those for carbonates are positively charged. Based on the contact angle measurements, it can be concluded that OF50-M nanoparticles have stronger interactions with the carbonate surface and hence more particles may have adsorbed than on the sandstone surfaces.

These observations show that the modified nanoparticles, OF50-M, have an ability to adsorb on various types of rock surfaces and alter the wettability from neutral- and oil-wet to water-wet conditions, which is an important criterion for more oil recovery from oil-wet petroleum formations.

4. Conclusions

Commercial silica nanoparticles were successfully surface modified to improve the colloidal stability of their dispersions under high salinity and high temperature conditions. Static (visual and turbidity) and accelerated stability analyses were performed at high temperatures to predict the long-term thermal stability. The stability tests of unmodified, control, and modified nanodispersions showed that only silane or only GLYMO modification is not sufficient to stabilize these nanofluids, in particular, at pH 6 and 7 and surface modification with combined SBS and GLYMO is necessary. The results of static stability analysis by Turbiscan^{TM} and visual method showed excellent thermal stability in 3.5% NaCl brine and SSW at 60 °C for a week and month, respectively. The accelerated stability analysis predicted that the modified nanoparticles could remain stable at least up to 6 months in the pH range of 4–7. The results of contact angle measurements showed that the modified nanoparticles were able to adsorb on neutral-wet rock surfaces and alter wettability to water-wet. A larger change in contact angle for carbonate surface than in sandstone surface showed that these particles could be more effective in carbonate reservoirs or reservoirs with high carbonate content and help improve oil recovery.

Author Contributions: N.J.H., conceptualization, methodology, data analysis, investigation and validation; Y.H.N., synthesis and nanoparticle modification; N.J.H. and Y.H.N., manuscript preparation; L.P.S., project lead; L.P.S. and O.T., review and editing; All authors have read and agreed to the published version of the manuscript.

Funding: This research was funded by Agency for Science, Technology, and Research (A*STAR), Singapore under IAF-PP Programme (project title: Advanced Functional Polymer Particle Technologies for Oil and Gas Industry; Grant No. A18B4a0094).

Acknowledgments: This project was funded by the Agency for Science, Technology and Research (A*STAR), Singapore under IAF-PP Programme (Project title: Advanced Functional Polymer Particle Technologies for the Oil and Gas Industry; Grant no.: A18B4a0094). The authors gratefully acknowledge the help provided by the Central Laboratory Facility for FTIR.

Conflicts of Interest: The authors declare no conflict of interest.

References

1. Green, D.W.; Willhite, G.P. *Enhanced Oil Recovery*, 2nd ed.; SPE Textbook Series; SPE: Richardson, TX, USA, 2018; Volume 6, pp. 1–2.
2. Mokheimer, E.M.A.; Hamdy, M.; Abubakar, Z.; Shakeel, M.R.; Habib, M.A.; Mahmoud, M. A comprehensive review of thermal enhanced oil recovery: Techniques Evaluation. *J. Energy Resour. Technol.* **2019**, *141*, 030801. [CrossRef]

3. Davarpanah, A. Parametric study of polymer-nanoparticles-assisted injectivity performance for axisymmetric two-phase flow in EOR processes. *Nanomaterials* **2020**, *10*, 1818. [CrossRef]

4. Massarweh, O.; Abushaikha, A.S. The use of surfactants in enhanced oil recovery: A review of recent advances. *Energy Rep.* **2020**, *6*, 3150–3178. [CrossRef]

5. Esmaeili, A.; Patel, R.B.; Singh, B.P. Applications of Nanotechnology in Oil and Gas Industry. *AIP Conf. Proc.* **2011**, *1414*, 133–136. [CrossRef]

6. Alsaba, M.T.; Al Dushaishi, M.F.; Abbas, A.K. A comprehensive review of nanoparticles applications in the oil and gas industry. *J. Pet. Explor. Prod Technol.* **2020**, *10*, 1389–1399. [CrossRef]

7. Fakoya, M.F.; Shah, S.N. Emergence of nanotechnology in the oil and gas industry: Emphasis on the application of silica nanoparticles. *Petroleum* **2017**, *3*, 391–405. [CrossRef]

8. Lau, H.C.; Yu, M.; Nguyen, U.P. Nanotechnology for oilfield applications: Challenges and opportunities. *J. Petrol. Sci. Eng.* **2017**, *157*, 1160–1169. [CrossRef]

9. Ko, S.; Huh, C. Use of nanoparticles for oil production applications. *J. Petrol. Sci. Eng.* **2019**, *172*, 97–114. [CrossRef]

10. Zhu, D.; Han, Y.; Zhang, J.; Li, X.; Feng, Y. Enhancing rheological properties of hydrophobically associative polyacrylamide aqueous solutions by hybriding with silica nanoparticles. *J. Appl. Polym. Sci.* **2014**, *131*, 40876. [CrossRef]

11. Sharma, T.; Iglauer, S.; Sangwai, J.S. Silica Nanofluids in an Oilfield Polymer Polyacrylamide: Interfacial Properties, Wettability Alteration, and Applications for Chemical Enhanced Oil Recovery. *Ind. Eng. Chem. Res.* **2016**, *55*, 12387–12397. [CrossRef]

12. Almahood, M.; Bai, B. The synergistic effects of NP-surfactant nanofluids in EOR applications. *J. Pet. Sci. Eng.* **2018**, *171*, 196–210. [CrossRef]

13. Wu, Y.; Chen, W.; Dai, C.; Huang, Y.; Li, H.; Zhao, M.; He, L.; Jiao, B. Reducing surfactant adsorption on rock by silica nanoparticles for enhanced oil recovery. *J. Pet. Sci. Eng.* **2017**, *153*, 283–287. [CrossRef]

14. Xu, D.; BBai Ziyu, M.; Qiong, Z.; Zhe, L.; Yao, L.; Hariong, W.; Jirui, H.; Wanli, K. A Novel Ultra-Low Interfacial Tension Nanofluid for Enhanced Oil Recovery in Super-Low Permeability Reservoirs. In Proceedings of the Paper SPE-192113-MS in SPE Asia Pacific Oil and Gas Conference and Exhibition, Brisbane, Australia, 23–25 October 2018. [CrossRef]

15. Bayet, A.E.; Junin, R.; Samsuri, A.; Piroozian, A.; Hokmabadi, M. Impact of metal oxide nanoparticles on enhanced oil recovery from limestone media at several temperatures. *Energy Fuels* **2014**, *28*, 6255–6266.

16. Maghzi, A.; Mohebbi, A.; Kharrat, R.; Ghazanfari, M.H. Pore-Scale Monitoring of Wettability Alteration by Silica Nanoparticles During Polymer Flooding to Heavy Oil in a Five-spot Glass Micromodel. *Transp. Porous Med.* **2011**, *87*, 653–664. [CrossRef]

17. Al-Anssari, S.; Barifcani, A.; Wang, S.; Maxim, L.; Iglauer, S. Wettability alteration of oil-wet carbonate by silica nanofluid. *J. Colloid Interface Sci.* **2016**, *461*, 435–442. [CrossRef]

18. Kanj, M.; Sakthivel, S.; Emmanuel, G. Wettability Alteration in Carbonate Reservoirs by Carbon Nanofluids. *Colloids Surf. A Physicochem. Eng. Asp.* **2020**, *598*, 124819. [CrossRef]

19. Binks, B.P.; Lumsdon, S.O. Stability of oil-in-water emulsions stabilised by silica particles. *Phys. Chem. Chem. Phys.* **1999**, *1*, 3007–3016. [CrossRef]

20. Chevalier, Y.; Bolzinger, M. Emulsions stabilized with solid nanoparticles: Pickering emulsions. *Colloids Surf. A Physicochem. Eng. Asp.* **2013**, *439*, 23–34. [CrossRef]

21. Kim, I.; Worthen, A.J.; Lotfollahi, M.; Johnston, K.P.; DiCarlo, D.A.; Huh, C. Nanoparticle-Stabilized Emulsions for Improved Mobility Control for Adverse-mobility Waterflooding. *Soc. Pet. Eng.* **2016**. [CrossRef]

22. Arab, D.; Kantzas, A.; Bryant, S.L. Nanoparticle stabilized oil in water emulsions: A critical review. *J. Pet. Sci. Eng.* **2018**, *163*, 217–242. [CrossRef]

23. Singh, R.; Misra, V. Stabilization of Zero-Valent Iron Nanoparticles: Role of Polymers and surfactants. In *Handbook of Nanoparticles*; Aliofkhazraei, M., Ed.; Springer: Cham, Switzerland, 2015. [CrossRef]

24. Worthen, A.J.; Tran, V.; Cornell, K.A.; Truskett, T.M.; Johnston, K.P. Steric stabilization of nanoparticles with grafted low molecular weight ligands in highly concentrated brines including divalent ions. *Soft Matter* **2016**, *12*, 2025–2039. [CrossRef]

25. Jang, H.; Lee, W.; Lee, J. Nanoparticle dispersion with surface-modified silica nanoparticles and its effect on the wettability alteration of carbonate rocks. *Colloids Surf. A Physicochem. Eng. Asp.* **2018**, *554*, 261–271. [CrossRef]

26. Griffith, C.; Daigle, H. Manipulation of Pickering emulsion rheology using hydrophilically modified silica nanoparticles in brine. *J. Colloid Int. Sci.* **2018**, *509*, 132–139. [CrossRef] [PubMed]

27. Griffith, C.; Daigle, H. A comparison of the static and dynamic stability of Pickering emulsions. *Colloids Surf. A Physicochem. Eng. Asp.* **2020**, *586*, 124256. [CrossRef]

28. Jadhunandan, P.P.; Morrow, N.R. Effect of Wettability on Waterflood Recovery for Crude-Oil/Brine/Rock Systems. *SPE Res. Eng.* **1995**, *10*, 40–46. [CrossRef]

29. Austad, T.; Milter, J. Spontaneous Imbibition of Water Into Low-Permeability Chalk at Different Wettabilities Using Surfactants. In Proceedings of the Paper SPE 37236 Presented at the SPE International Symposium on Oilfield Chemistry, Houston, TX, USA, 18–21 February 1997.

30. Standnes, D.C.; Austad, T. Nontoxic Low-cost Amines as Wettability Alteration Chemicals in Carbonates. *J. Pet. Sci. Eng.* **2003**, *39*, 431. [CrossRef]

31. Spinler, E.A.; Zornes, D.R.; Tobola, D.P.; Moradi-Araghi, A. Enhancement of Oil Recovery Using a Low Concentration of Surfactant to Improve Spontaneous and Forced Imbibition in Chalk. In Proceedings of the Paper SPE 59290 Presented at the SPE/DOE Improved Oil Recovery Symposium, Tulsa, OK, USA, 3–5 April 2000.
32. Sharma, G.; Mohanty, K. Wettability Alteration in High-Temperature and High-Salinity Carbonate Reservoirs. *SPE J.* **2013**, *18*, 646–655. [CrossRef]
33. Jarrahian, K.; Seiedi, O.; Sheykhan, M.; Sefti, M.V.; Ayatollahi, S. Wettability alteration of carbonate rocks by surfactants: A mechanistic study. *Colloids Surf. A Physicochem. Eng. Asp.* **2012**, *410*, 1–10. [CrossRef]
34. Al-Yaseri, A.Z.; Lebedev, M.; Vogt, S.J.; Johns, M.L.; Barifcani, A.; Iglauer, S. Pore-scale analysis of formation damage in Bentheimer sandstone with in-situ NMR and micro-computed tomography experiments. *J. Pet. Sci. Eng.* **2015**, *129*, 48–57. [CrossRef]
35. Estephan, Z.G.; Jaber, J.A.; Schlenoff, J.B. Zwitterion-stabilized silica nanoparticles: Toward nonstick nano. *Langmuir* **2010**, *26*, 16884–16889. [CrossRef]
36. LUM GmbH. *SepView User Manual*; LUM GmbH: Berlin, Germany, 2016.
37. Worthen, A.J.; Alzobaidi, S.; Tran, V.; Iqbal, M.; Liu, J.S.; Cornell, K.A.; Kim, I.; DiCarlo, D.A.; Bryant, S.L.; Huh, C.; et al. Design of Nanoparticles for Generation and Stabilization of CO_2-in-Brine Foams with or without Added Surfactants. 2019. Available online: https://arxiv.org/abs/1811.11217v2Website (accessed on 22 February 2021).

 *nanomaterials*

Article

Pore- and Core-Scale Insights of Nanoparticle-Stabilized Foam for CO$_2$-Enhanced Oil Recovery

Zachary Paul Alcorn [1,*], **Tore Føyen** [1,2], **Jarand Gauteplass** [1], **Benyamine Benali** [1], **Aleksandra Soyke** [1] and **Martin Fernø** [1]

[1]   Department of Physics and Technology, University of Bergen, 5007 Bergen, Norway; tore.foyen@uib.no (T.F.);
      Jarand.Gauteplass@uib.no (J.G.); Benyamine.benali@uib.no (B.B.); aleksandra.soyke@uib.no (A.S.);
      Martin.Ferno@uib.no (M.F.)
[2]   SINTEF Industry, 7034 Trondheim, Norway
*   Correspondence: zachary.alcorn@uib.no

Received: 4 September 2020; Accepted: 17 September 2020; Published: 25 September 2020

Abstract: Nanoparticles have gained attention for increasing the stability of surfactant-based foams during CO$_2$ foam-enhanced oil recovery (EOR) and CO$_2$ storage. However, the behavior and displacement mechanisms of hybrid nanoparticle–surfactant foam formulations at reservoir conditions are not well understood. This work presents a pore- to core-scale characterization of hybrid nanoparticle–surfactant foaming solutions for CO$_2$ EOR and the associated CO$_2$ storage. The primary objective was to identify the dominant foam generation mechanisms and determine the role of nanoparticles for stabilizing CO$_2$ foam and reducing CO$_2$ mobility. In addition, we shed light on the influence of oil on foam generation and stability. We present pore- and core-scale experimental results, in the absence and presence of oil, comparing the hybrid foaming solution to foam stabilized by only surfactants or nanoparticles. Snap-off was identified as the primary foam generation mechanism in high-pressure micromodels with secondary foam generation by leave behind. During continuous CO$_2$ injection, gas channels developed through the foam and the texture coarsened. In the absence of oil, including nanoparticles in the surfactant-laden foaming solutions did not result in a more stable foam or clearly affect the apparent viscosity of the foam. Foaming solutions containing only nanoparticles generated little to no foam, highlighting the dominance of surfactant as the main foam generator. In addition, foam generation and strength were not sensitive to nanoparticle concentration when used together with the selected surfactant. In experiments with oil at miscible conditions, foam was readily generated using all the tested foaming solutions. Core-scale foam-apparent viscosities with oil were nearly three times as high as experiments without oil present due to the development of stable oil/water emulsions and their combined effect with foam for reducing CO$_2$ mobility

Keywords: nanoparticles; foam; CO$_2$ EOR; CO$_2$ mobility control

---

1. Introduction

An energy transition to a net-zero society is a global challenge in need of affordable, low-risk technologies. Carbon capture, utilization and storage (CCUS) is a crucial technology for substantial emission cuts for many energy-intensive industries to achieve the ambitious climate goals of the Paris Agreement [1]. CCUS involves capturing CO$_2$ from industrial sources and injecting it into subsurface reservoirs for simultaneous storage and energy production, via CO$_2$-enhanced oil recovery (EOR). Permanent CO$_2$ storage coupled with CO$_2$ EOR can provide affordable and reliable energy for our developing world while reducing the life-cycle carbon emissions of fossil fuels.

CO_2 EOR has been developed and widely implemented over the past 50 years. CO_2 is an excellent solvent in EOR processes because it is miscible with most crude oils at reservoir conditions. Above miscibility conditions, CO_2 swells the oil and reduces its viscosity resulting in increased recovery. Laboratory corefloods have reported high microscopic displacement efficiency and oil recoveries of nearly 100% [2]. However, field-scale operations often report lower than expected recoveries due to poor sweep efficiency and high CO_2 mobility [3,4]. These issues stem from reservoir heterogeneity and the low viscosity and density of CO_2 compared to reservoir fluids.

CO_2 foam can mitigate the impacts of high CO_2 mobility and reservoir heterogeneity by effectively increasing CO_2 viscosity, reducing its relative permeability and diverting CO_2 flow from high permeability zones [5]. CO_2 foam is generated in porous media by injecting foaming solution with CO_2, either simultaneously or in alternating slugs. The foam is a dispersion of CO_2 in liquid where stable liquid films, called lamellae, block some of the pathways for CO_2 flow [6]. Lamellae are commonly stabilized by surfactants. However, surfactant-stabilized foams can break down in the reservoir due to surfactant adsorption, the presence of oil, and at elevated temperatures and salinities. Therefore, their ability to reduce CO_2 mobility can be limited. The addition of silica nanoparticles to the surfactant-stabilized CO_2 foam has been shown to increase the strength and stability of the foam system and provide increased oil recovery [7,8].

Spherical silica nanoparticles are the most commonly used for EOR applications [9]. They are particles with a size up to 100 nm with intrinsic properties different from those found in the bulk of the material due to their high surface-to-volume ratio. Stable emulsions are generated using nanoparticles because a rigid monolayer is formed on the droplet surface and the particles are irreversibly attached to the interface. These emulsions may withstand high-temperature reservoir conditions without agglomeration and the nanoparticles may be further surface-treated to improve stability in harsh conditions. In addition, the small size of the particles, two orders of magnitude smaller than colloidal particles, make them suitable for flow through small pore throats in rock [10,11].

Whether stabilized by surfactants, nanoparticles, or a combination of both, bulk foams are typically composed of bubbles smaller than the containers they are within whereas foam in porous media is composed of bubbles about the same size or larger as the pore space [12]. For foam to generate, lamella creation must exceed lamella destruction. Capillary forces dominate lamella creation by three main mechanisms: leave behind, snap-off and lamella division [5,13].

An issue with foam for EOR applications is the impact of oil on foam (lamellae) stability. Many studies report that oil hinders foam generation and can destabilize already generated foam [14–16]. However, these findings are mostly based upon bulk tests at immiscible conditions with surfactant-stabilized foam, which may not necessarily represent foam in porous media and at miscible conditions for CO_2 and oil. In any case, foam behavior in the presence of oil involves several interactions between the foam, oil, and rock, which may be either detrimental or beneficial to the foam process [17,18]. These interactions include emulsification–imbibition, pseudo emulsions, and entering and spreading [19,20].

In the absence of oil, foam coalescence can reduce the number of bubbles by two mechanisms: texture (bubble size) coarsening by diffusion, often referred to as Ostwald ripening, or capillary suction drainage [21]. Diffusion occurs by the transport of gas from smaller bubbles to larger bubbles, with lower internal pressure, which results in fewer bubbles [22,23]. Capillary suction drainage occurs when the water saturation approaches a saturation value where the lamellae are no longer stable, as the capillary pressure exceeds the maximum disjoining pressure of the foam film and drains the lamellae [24,25].

The majority of earlier work has focused on foam generation and the coalescence of surfactant-stabilized CO_2 foams in the absence and presence of oil at immiscible conditions. However, much less is known about the role of nanoparticles in the absence and presence of oil at miscible conditions. Thus, this study aimed to thoroughly characterize the dominant foam generation mechanisms and determine the role of nanoparticles for stabilizing CO_2 foam and reducing CO_2

mobility. In addition, we shed light on the influence of oil on foam generation and stability. We present a pore- to core-scale characterization of hybrid nanoparticle–surfactant foam formulation for CO_2 mobility control for CO_2 EOR and CO_2 storage. Experimental results compared the hybrid foaming solution to foam stabilized by only surfactant or nanoparticles, in the presence and absence of oil.

2. Materials and Procedures

2.1. Pore-Scale System

Two foaming agents were used to study foam generation, stability and coalescence. One was a nonionic surfactant (Huntsman *Surfonic L24-22*, Houston, TX, USA), a linear ethoxylated alcohol. The other foaming agent was a surface-modified spherical silica nanoparticle (Nouryon *Levasil CC301*, Amsterdam, The Netherlands). Foaming solutions were made by dissolving each foaming agent, either separately or combined, in 35,000 ppm NaCl brine at the concentrations shown in Table 1. CO_2 with 99.999% purity was used. The pore space was cleaned between injection cycles using 2-proponal-water azeotrope (IPA). For experiments in the presence of oil, a refined oil (n-Decane, $C_{10}H_{22}$) was used to obtain first-contact miscibility with CO_2.

Table 1. Composition of the foaming solutions used in pore- and core-scale experiments.

Foaming Agents	Concentration, *Component*	Scale
Nanoparticle (NP)	1500 ppm, *Levasil CC301*	Pore
Surfactant (SF)	3500 ppm, *Surfonic L24-22*	Core
	5000 ppm, *Surfonic L24-22*	Pore and Core
Hybrid (SF + NP)	3500 ppm, *Surfonic L24-22* + 1500 ppm, *Levasil CC301*	
	5000 ppm, *Surfonic L24-22* + 1500 ppm, *Levasil CC301*	Pore
	5000 ppm, *Surfonic L24-22* + 150 ppm, *Levasil CC301*	
	3500 ppm, *Surfonic L24-22* + 150 ppm, *Levasil CC301*	Core

The micromodel was composed of a rectangular etched silicon wafer with an irregular porous structure bonded to a transparent borosilicate glass with dimensions of 26.96 mm × 22.50 mm (Figure 1) and a constant etching depth of 30 μm. The pore pattern was a simplified two-dimensional projection of real pore structures with connected pores that allow flow with discontinuous, irregularly shaped grains that provide tortuosity. The chemical composition of the crystalline silicon and borosilicate glass are similar to sandstone and are chemically inert to the injected fluids. Complete manufacturing procedures can be found elsewhere [26,27].

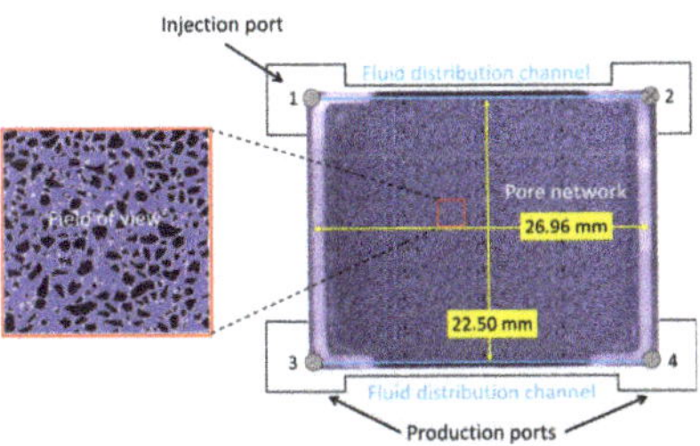

Figure 1. Dimensions of the micromodel, location of the flow ports and the fluid distribution channels. The focused field of view is shown on the left. Injection was into port 1 and production was from ports 3 and 4. Port 2 was closed. The entire pore network consisted of 36 repetitions of a single 749-grain pore pattern. The grain size distribution ranged from 100 to 79,000 μm^2 and the pore throat distribution ranged from 10 to 200 μm. The average pore throat length was 89 μm.

The micromodel had a porosity of 61%, permeability of 3000 mD and pore volume (PV) of 11.1 μL. The porous pattern (27,000 grains) had 36 (4 × 9) repetitions of a pore network with 749 unique grains. The grain size distribution of the 749-grain pattern ranged between 100 and 79,000 μm² and the pore throat width distribution ranged from 10 to 200 μm. Flow ports were located at each corner of the micromodel with the inlet at ports 1 and 2 and the outlet at ports 3 and 4. The micromodel was positioned in the bottom part of a two-piece polyether ether ketone (PEEK) plastic micromodel holder. The top part had an open window for direct visual observation. The micromodel holder was placed on a motorized stage below a microscope (Axio Zoom. V16, Zeiss, Jena, Germany). The microscope software controlled the zoom, focus, illuminator intensity, imaging, and the motorized stage. Additional details on the micromodel set-up can be found in [28].

2.2. Pore-Scale Procedure

The micromodel system was pressurized to 100 bar using a backpressure system at 25 °C for experiments in the absence and presence of oil. For experiments in the absence of oil, foaming solution was first injected to completely saturate the micromodel before injecting dense (liquid) phase CO_2 at a constant volumetric flow rate of 4 μL/min. The foaming solutions consisted of 1500 ppm nanoparticles, 5000 ppm surfactant, and two hybrid solutions with 5000 ppm surfactant combined with 1500 ppm or 150 ppm nanoparticles. An overview of the foaming solutions are listed in Table 1. A baseline, without foaming solution, was also conducted for comparison. For experiments in the presence of oil, the micromodel was initially saturated with distilled water before injecting six pore volumes of oil. Distilled water was then injected for an additional six pore volumes to achieve residual oil saturation. The micromodel was then saturated with the hybrid 3500 ppm surfactant and 1500 ppm nanoparticle foaming solution before CO_2 injection began at a constant rate of 1 μL/min. For all experiments, CO_2 was injected in port 1 (inlet), port 2 was closed and ports 3 and 4 (outlet) were open and kept at 100 bar using the backpressure system (Figure 1). The microscope settings (light intensity, aperture, and shutter time) were optimized for image processing and remained constant. Images were acquired of the entire micromodel with high spatial resolution (4.38 μm/pixel) by stitching multiple overlapping images. The image acquisition time of the porous pattern (121 separate images) was 73 s. A focused field of view was selected, which was representative of the remainder of the micromodel, for detailed analysis and to minimize the capillary end effects. Raw images from the experiments show the grains as dark and opaque and the pore space in a grayish-blue hue. The gas/liquid interfaces (lamellae) were white due to the diffusive ring-illuminator of the microscope. Foam generation and coalescence were also analyzed by utilizing the Python Library OpenCV [29] to identify bubble number and size.

2.3. Core-Scale System

The core-scale experiments used the same brine as the pore-scale work. In experiments with only surfactant in the foaming solution, a 3500 ppm or 5000 ppm concentration was used. In experiments with the hybrid foaming solutions, a 3500 ppm surfactant concentration was used with either 1500 ppm or 150 ppm nanoparticles to evaluate the concentration sensitivity for foam stabilization. See Table 1 for an overview of the foaming solutions. A single outcrop Bentheimer sandstone core was used for all experiments to eliminate the impacts of variable core properties. The core was cleaned and dried before being 100% saturated with brine under vacuum. Porosity and pore volumes were calculated based on the weight differential before and after saturation. Absolute permeability was measured between each experiment by injecting brine until a stable differential pressure was obtained for three different injection rates. The permeability of the core was 1400 millidarcy with a porosity of 24% (Table 2).

Table 2. Core properties of the Bentheimer sandstone used in the experimental work.

Core Properties	Value
Length (cm)	24.6 ± 0.01
Diameter (cm)	3.64 ± 0.01
Pore Volume (mL)	68.23
Porosity	0.24
Permeability (mD)	1400

2.4. Core-Scale Procedure

The brine-saturated sandstone core was wrapped in a 0.1-mm thick nickel foil to reduce the radial CO_2 diffusion into the confinement oil before installation into the Viton rubber sleeve. The core was then mounted in a vertically oriented Hassler-type core holder and placed inside a heating cabinet. Experimental conditions were set to 40 °C and 200 bar with a net overburden pressure of 70 bar. At these conditions, CO_2 is supercritical and has a similar density as in the pore-scale experiments. A differential pressure transducer and two absolute pressure transducers monitored pressure response at the inlet and outlet. Figure 2 shows the experimental set-up, modified from [30].

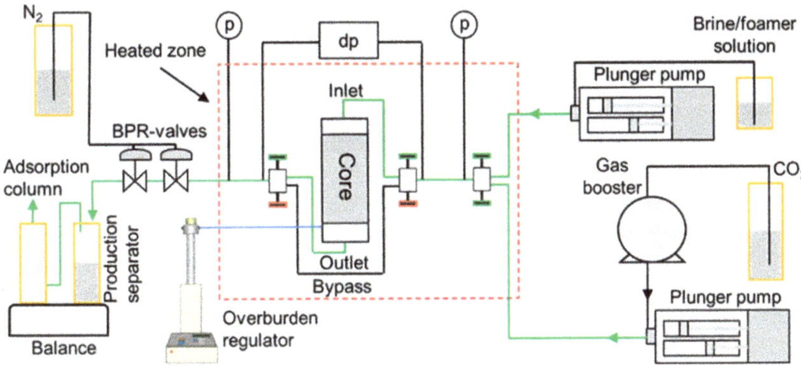

Figure 2. Experimental setup used for the core-scale foam experiments. Green lines indicate the fluid flow directions during the injection of CO_2 and the foaming solution. Pure CO_2 was pressurized by a gas booster and injected using a Quizix Q6000-10k plunger pump. Foaming solutions were injected using a Quizix Q5000-10k plunger pump. Injection was performed through a series of needle valves (marked green for open, red for closed) to the top of the core. Produced fluids were depressurized downstream through a series of backpressure regulator (BPR) valves and measured in the production separator and associated water adsorption column using a digital balance. Modified from [30].

Foam apparent viscosity is a measure of foam generation, strength and stability. An increase in apparent viscosity indicates a generation of foam and a higher value of apparent viscosity corresponds to a stronger foam. Foam apparent viscosity (μ_{app}) was quantified from the experimental superficial velocities and measured pressure drop [31] by

$$\mu_{app} = \frac{k\nabla p}{\left(u_l + u_g\right)} \tag{1}$$

where k is the absolute permeability of the porous media, ∇p is the measured pressure gradient and u_l and u_g are the superficial velocities of liquid and gas, respectively [32]. The effect of nanoparticles on foam strength and stability was evaluated by comparing dynamic experimental apparent viscosity results using foaming solutions with and without nanoparticles.

The injection scheme for the core-scale experiments in the absence of oil was adapted from [33]. First, a minimum of three PVs of foaming solution was injected to satisfy adsorption, displace the initial brine and fully saturate the pore space. Then, CO_2 was injected from the top of the vertically mounted core at a superficial velocity of 4 ft/day for approximately six PVs. Unsteady state apparent foam viscosities were calculated as a function of time (PVs injected) using Equation (1). A minimum of two experiments were performed for each individual foaming solution. A baseline experiment, without foaming solution, was also conducted for comparison. The core was cleaned between experiments by injecting solutions of IPA before being re-saturated with brine and then foaming solution.

The core-scale procedure in the presence of oil was developed to obtain approximately 30% residual oil before evaluating foam generation and stability. First, a primary drainage with n-Decane for nearly one PV was conducted followed by a waterflood for one PV. Foaming solution was then injected for at least three PVs at a low and high rate. Finally, CO_2 was continuously injected at 4 ft/day for 10 to 14 PVs. A minimum of two experiments was performed for each individual foaming solution.

3. Results and Discussion

3.1. Pore-Scale: Foam in the Absence of Oil

Figure 3 shows pore-scale images from four experiments with different foaming solutions. Three time steps are shown which correspond to pre-foam generation (PV = 1.3), peak foam generation and post-foam generation (PV = 20.1). The images show a focused field of view with CO_2 injection from the top to the bottom for each image. The dark opaque areas are grains, the grayish-blue open areas are the pore space and the thin white films are lamellae.

The experiment with only nanoparticles present (1500 NP) generated weak foam as indicated by the continuous distribution of open flow paths and very few lamellae or bubbles (Figure 3, left column). Thus, CO_2 mobility remained high and was comparable to the baseline without any foaming agent. CO_2 injection with the three surfactant-laden foaming solutions resulted in the generation of densely distributed, finely textured foam, which significantly reduced CO_2 mobility during the peak foam generation stage (5000 SF, 5000 SF + 1500 NP and 5000 SF + 150 NP). Individual bubbles were located near the ends of pore throats and several bubbles filled individual pore bodies, suggesting snap-off as the primary foam generation mechanism. Because the pore bodies had a larger area than the pore throats, repeated snap-off occurred until the pore body was filled with bubbles, a phenomenon also described by [34]. Dynamic observations also revealed many individual lamellae spanning across pore throats. These lamellae may have formed from the leave-behind mechanism because CO_2 was injected into a surfactant saturated porous media in a drainage-like process. The rise in capillary pressure during drainage can cause lamellae generation by both leave-behind and snap-off as gas enters the pore network [35].

Direct visual observations of the experiment with the hybrid foaming solution containing 5000 ppm surfactant and 1500 ppm nanoparticles revealed a continuous open flow path for CO_2 throughout the duration of the experiment (Figure 3, red line, 5000 SF + 1500 NP). No lamellae impeded CO_2 flow in this region and the CO_2 relative permeability was reduced by the presence of lamellae in the remainder of the pore network. Therefore, within this focused field of view, a continuous gas-foam was generated.

Foaming Solution

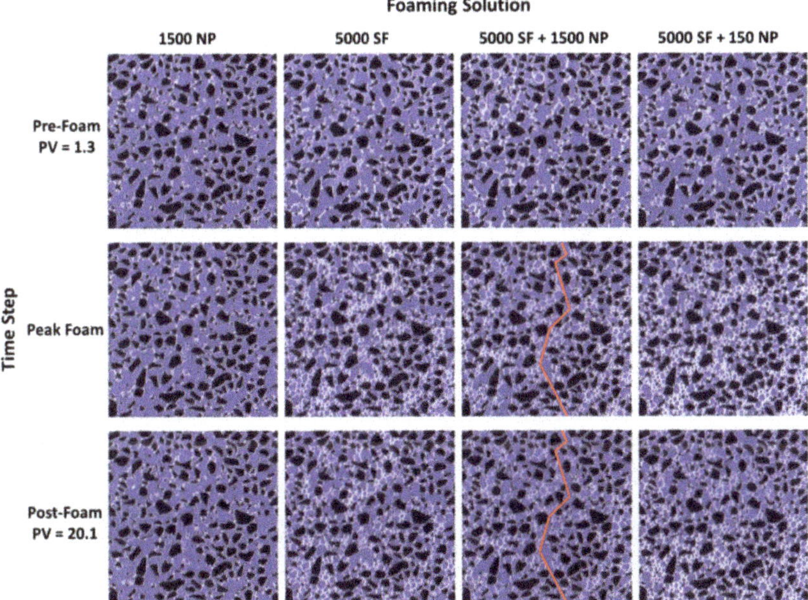

Figure 3. The pore-scale images of a focused field of view during the injection of dense phase CO_2 into a micromodel saturated with four different foaming solutions at 100 bar and 25 °C. Experiments with different foaming solutions are shown across the top: 1500 ppm nanoparticles (1500 NP), 5000 ppm surfactant (5000 SF), hybrid 5000 ppm surfactant and 1500 ppm nanoparticles (5000 SF + 1500 NP) and hybrid 5000 ppm surfactant and 150 ppm nanoparticles (5000 SF + 150 NP). Injection was from top to bottom in each image. The dark opaque areas are grains, the grayish-blue open areas are the pore space and the thin white films are lamellae. Individual image dimensions are 2190 × 2190 μm. The grain size ranged from 100 to 79,000 μm^2 and the pore throat distribution ranged from 10 to 200 μm for the entire micromodel.

Figure 4 quantifies the number of bubbles versus the bubble size for the images shown in Figure 3. Bubble number and size were used as indications of foam generation and strength where a higher bubble number corresponded to a finer textured foam. All foaming solutions containing surfactant-generated small bubbles ($\leq 10^3$ μm^2) at the peak generation stage. In the post-foam generation stage, the total number of bubbles decreased and their size increased; hence, the foam texture coarsened, increasing CO_2 mobility as CO_2 was continuously injected. The hybrid foaming solutions with either 1500 ppm or 150 ppm nanoparticles showed similar behavior, indicating that foam strength and stability was not sensitive to nanoparticle concentration when used together with the selected surfactant.

Pore-scale foam behavior was also analyzed by examining the total bubble number (N_i) as a function of the PV of CO_2 injected. The number of bubbles during foam generation and coalescence (N_{bubble}) were normalized to baseline ($N_{baseline}$) for the four foaming solutions. Figure 5 shows the normalized bubble number as a function of PV injected for each foaming solution for the focused field of view. Foam generation (as indicated by bubble number) increased from approximately 9 to 11 times the baseline for all foaming solutions. Peak foam generation was reached after approximately seven PVs of the CO_2 injected. After peak foam generation, the number of bubbles steadily decreased from bubble coarsening as the dominant coalescence mechanism as observed in Figure 3. The hybrid foaming solutions, containing nanoparticles and surfactant, had a limited impact on the number of bubbles and foam stability during continuous CO_2 injection.

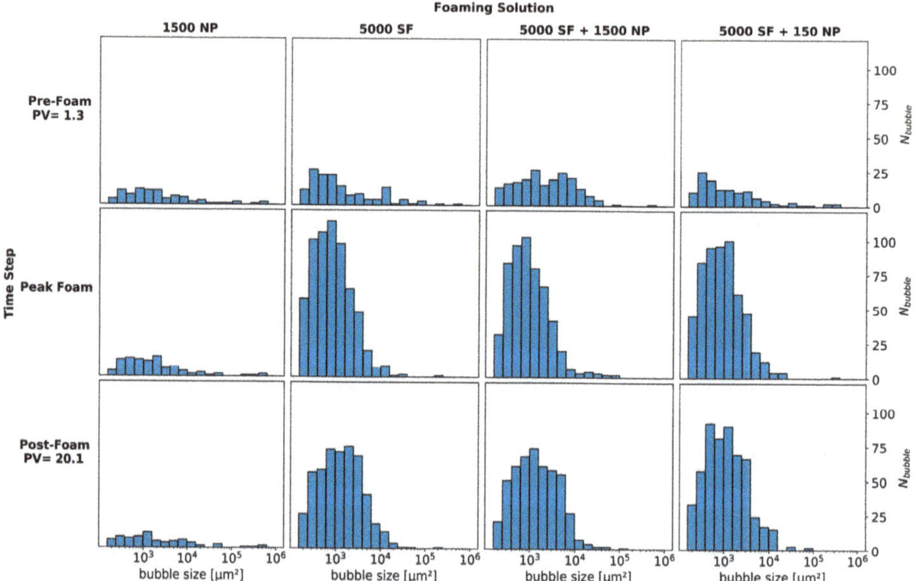

Figure 4. The number of bubbles (N_{bubble}) versus bubble size for the micromodel experiments with four different foaming solutions. Foaming solutions are shown across the top and include 1500 ppm nanoparticles (1500 NP), 5000 ppm surfactant (5000 SF), hybrid 5000 ppm surfactant and 1500 ppm nanoparticles (5000 SF + 1500 NP) and hybrid 5000 ppm surfactant and 150 ppm nanoparticles (5000 SF + 150 NP).

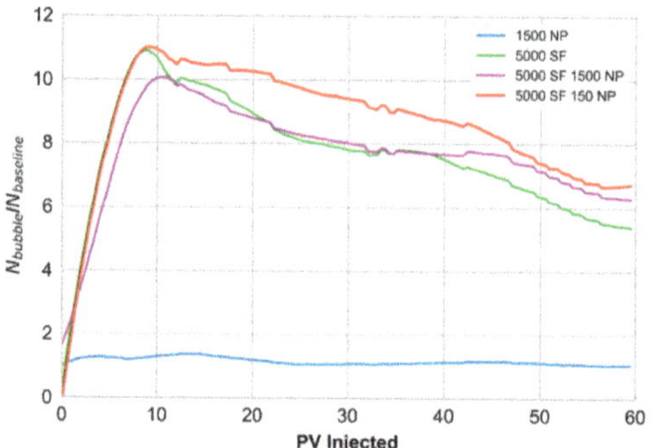

Figure 5. Development in normalized bubble number as a function of pore volume (PV) injected using four different foaming solutions for the focused field of view. The blue curve represents the foaming solution with 1500 ppm nanoparticles (1500 NP), the green curve represents the 5000 ppm surfactant solution (5000 SF), the purple curve represents the hybrid solution with 5000 ppm surfactant and 1500 ppm nanoparticles (5000 SF + 1500 NP) and the red curve represents the hybrid solution with 5000 ppm surfactant and 150 ppm nanoparticles (5000 SF + 150 NP).

The two-dimensional geometry of the micromodel likely resulted in multiple bubbles per pore because the widths of some of the pore throats were narrower than the pore throat depths. Therefore, pore-scale foam texture may not have a direct relation to foam in three-dimensional porous media. Many studies report that in situ foam usually consists of bubbles about the same size or larger than pore bodies based upon effluent analysis during laboratory experiments and the large flow resistance for bubbles smaller than pores [12,36,37]. In addition, most mechanistic foam models [38–40] assume a single bubble per pore and that discrete bubbles flow through the porous media, where foam strength is controlled by foam texture (bubble size). The latter assumptions are supported by the pore-scale observations reported here.

3.2. Pore-Scale: Foam in the Presence of Oil

Dynamic foam generation in the presence of oil was evaluated by injecting CO_2 into a micromodel saturated with a hybrid foaming solution and oil. The aim was to evaluate the impact of oil on foam generation and gain insight on the influence of oil/water emulsions during CO_2 foam processes. Figure 6 shows the pore-scale images of the unsteady-state CO_2 injection in the presence of oil with the hybrid foaming solution containing 3500 ppm surfactant and 1500 ppm nanoparticles. Three stages of the experiment are shown which correspond to before CO_2 injection, the start of CO_2 injection, and during CO_2 injection. Each image was acquired with 75 s between each time step.

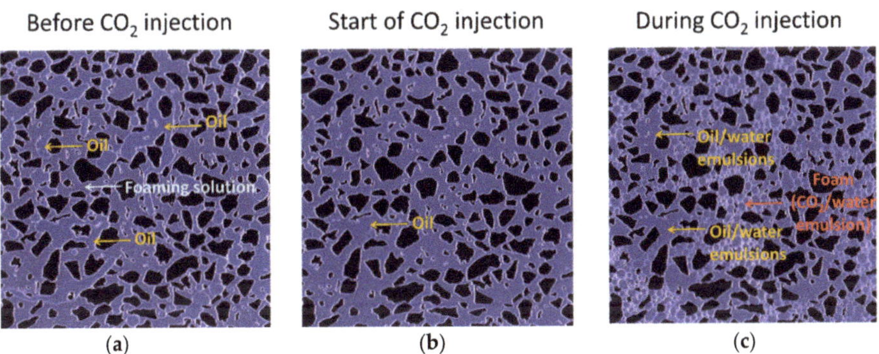

Figure 6. Pore-scale images of a focused field of view during the injection of dense phase CO_2 into a micromodel saturated with a hybrid foaming solution and oil at 100 bar and 25 °C. Three stages of the experiment are shown which correspond to: (**a**) before CO_2 injection; (**b**) the start of CO_2 injection; and (**c**) during CO_2 injection. Injection was from top to bottom in each image. The dark opaque areas are grains, the grayish-blue open areas are the pore space filled and the thin white films are the lamellae. Individual image dimensions are 2190×2190 μm. The grain size ranged from 100 to 79,000 μm^2 and the pore throat distribution ranged from 10 to 200 μm for the entire micromodel.

Before CO_2 injection, the micromodel was initially saturated with foaming solution and oil (Figure 6a). Foaming solution appears as the continuous liquid phase, whereas oil is seen as isolated globules in interconnected pores. At the start of CO_2 injection (Figure 6b), the oil globules faded due to miscibility between CO_2 and oil. As CO_2 injection continued, the oil was displaced by CO_2 and foam readily generated in areas where oil was not present. Oil not displaced formed oil/water emulsions and occupied pores without foam present (Figure 6c). The foam (CO_2/water emulsion) had thicker lamellae compared to the oil/water emulsions likely due to interfacial tension differences at these conditions as also observed in [41]. Compared to foam (CO_2/water emulsion) alone, the combined effect of oil/water emulsions and foam further reduced CO_2 mobility. This resulted in increased "foam" strength as also observed in the core-scale experiments in the presence of oil (discussed in Section 3.4).

3.3. Core-Scale: Foam in the Absence of Oil

Dynamic foam generation and stability in the absence of oil was evaluated by injecting CO_2 into cores saturated with different foaming solutions. This set of experiments established conditions to investigate foam behavior during prolonged periods of CO_2 injection in a drainage-like process. Figure 7a shows the apparent viscosity versus pore volume of CO_2 injected for the CO_2 foam stability scans with foaming solutions containing only surfactant at concentrations of 3500 ppm (green curves) and 5000 ppm (blue curves). Figure 7b shows the results from the experiments using the two hybrid foaming solutions with 3500 ppm surfactant and 150 ppm nanoparticles (orange curves) and 3500 ppm surfactant and 1500 ppm nanoparticles (red curves). A baseline scan, without foaming solution, is also shown in each figure for comparison (black curves).

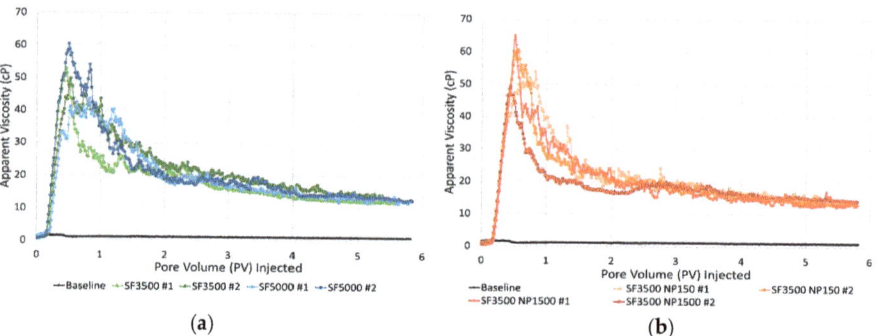

Figure 7. Apparent viscosity versus pore volume of the CO_2 injected for the unsteady state CO_2 injections into cores pre-saturated with foaming solutions containing: (**a**) 3500 ppm surfactant (green curves) and 5000 ppm surfactant (blue curves); (**b**) hybrid foaming solutions containing 3500 ppm surfactant and 150 ppm nanoparticles (orange curves) and 3500 ppm surfactant and 1500 ppm nanoparticles (red curves). The black curve is the baseline with only brine.

For all experiments, the rapid and linearly increasing apparent viscosity until 0.2 PV injected indicated that foam was generated as CO_2 invaded the core saturated with foaming solution. Apparent viscosity steadily increased, from 0.2 to 0.5 PV injected, as foam continued to generate and propagate into the core. A peak in apparent viscosity (foam strength) was achieved after approximately 0.5 PV was injected. The magnitude of the peak apparent viscosity varied from 45 to 65 cP for all experiments. The peak in apparent viscosity indicated a transition from a period of predominantly foam generation to predominantly foam coalescence. The development of a continuous CO_2 flow path not impeded by lamellae caused the foam to coalesce, likely related to a combination of bubble rupture and foam displacement. The CO_2 flow path rapidly reduced the apparent viscosity just before one PV was injected. After about six PVs were injected, the initial CO_2 viscosity was not fully recovered due to trapped bubbles in the pore space, which continued to reduce CO_2 mobility.

The difference in dynamic foam generation and coalescence processes for the foaming solutions with and without nanoparticles were insignificant. Including nanoparticles in the surfactant-laden foaming solution did not result in a more stable foam and the type of foaming solutions did not clearly affect the apparent viscosity of the foam. Therefore, the surfactant contributed mostly to foam generation and the nanoparticles had only minor impacts on the foam strength and stability in these experiments. The experiments with the hybrid foaming solutions (Figure 7b) revealed similar foam behavior independent of nanoparticle concentration. Despite an order of magnitude difference in nanoparticle concentration, the measured apparent viscosities and stability of the foam were similar. Thus, the nanoparticle concentrations of 150 ppm gave similar performance as the nanoparticle concentrations of 1500 ppm when used with the selected surfactant. The next set of experiments focused on evaluating the same foaming solutions in the presence of oil, a condition known to destabilize some surfactant-based foams.

3.4. Core-Scale: Foam in the Presence of Oil

Dynamic foam generation and stability for foaming solutions with and without nanoparticles in the presence of oil was evaluated by injecting CO_2 into a core saturated with each foaming solution. The core contained a residual oil saturation of around 30% prior to being flooded with foaming solution and then CO_2. Each experiment was conducted a minimum of two times for reproducibility. Figure 8 shows the average apparent viscosity (cP) versus the pore volume of CO_2 injected for the unsteady state CO_2 foam stability scans in the presence of oil. Experiments using the foaming solution with only surfactant are shown with the blue curve and experiments with the hybrid foaming solution are shown with the red curve.

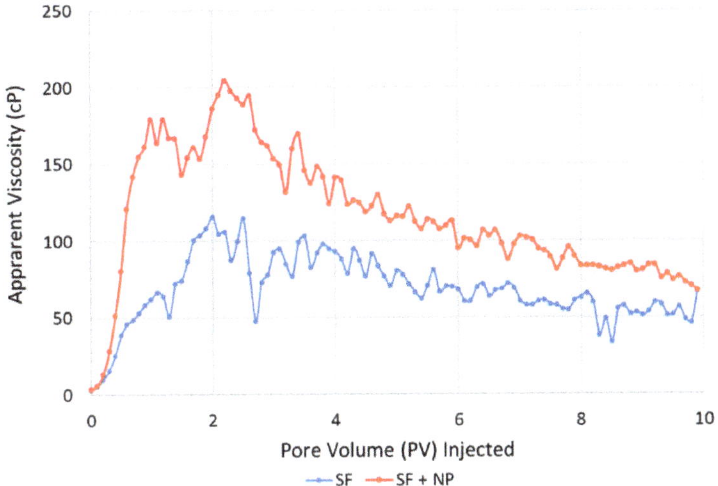

Figure 8. Apparent viscosity versus pore volume of the CO_2 injected for the unsteady state CO_2 injections in cores with residual oil (S_{or}) and pre-saturated with a hybrid foaming solution containing surfactant and nanoparticles (SF + NP, red curve) or a foaming solution containing only surfactant (SF, blue curve).

Both types of foaming solutions generated foam within the first 0.2 PV injected. However, the hybrid foaming solution generated foam more rapidly (faster increase in apparent viscosity) than the solution containing only surfactant. In addition, the hybrid foaming formulation generated a stronger (higher apparent viscosity) foam, compared to the solution containing only surfactant. The increased apparent viscosity for both types of solution indicated that each formulation generated foam with the residual oil present.

Apparent foam viscosity values with the hybrid solution in the presence of oil (Figure 8, red curve) were nearly three times as high as the experiments without oil present (Figure 7b). In the presence of oil, the foaming solution with only surfactant (Figure 8, blue curve) had foam-apparent viscosity values about twice as high as experiments in the absence of oil (Figure 7a). This is related to the development of oil/water emulsions, which were likely stabilized by each respective foaming agent. The emulsions influenced the calculated apparent viscosities (differential pressure) and are indistinguishable from foam (CO_2/water emulsion). Nonetheless, the oil/water emulsions highlight an important facet of the CO_2 foam process, which can be beneficial to enhancing oil recovery by increasing the capillary number (increased viscous forces and lower interfacial tension) [42].

3.5. From Pore- to Core-Scale

The similarity in foam generation and coalescence during unsteady-state CO_2 injections at the pore- and core-scale is striking. Figures 5 and 7 reveal dynamic foam generation and coalescence processes with similar behavior at two different length scales. The experiments in this work were characterized by a period of rapid foam generation during drainage-like CO_2 injection and a period of foam coalescence during prolonged CO_2 injection. The decline in foam strength, at both scales, was related to the development of open CO_2 flow paths through the generated foam. This phenomenon was a result of bubble coarsening from diffusion. The pore-scale observations unlocked real-time insights on in situ foam behavior that may help explain the observations from the core-scale experiments. Since foam was rapidly generated at both scales (due to ideal conditions for foam generation), the coalescence mechanisms during continued CO_2 injection at the pore-scale may be applied at the core-scale with some level of confidence. It is understood that foam will dry out as more CO_2 is injected and not supplemented with additional surfactant solution. Here, we showed one of the physical mechanisms responsible for such behavior.

In addition, the experiments in the presence of oil revealed the importance of stable oil/water emulsions on the CO_2 foam process. The insights from pore-scale experiments with oil shed light on the influence of oil/water and CO_2/water emulsions on CO_2 mobility reduction. Higher foam apparent viscosities were calculated for the core-scale experiments with oil present and were likely related to the development of the oil/water emulsions. Because apparent viscosity is used as an indication of foam generation and strength in laboratory experiments, care must be taken when interpreting the results from coreflood studies with the presence of stable oil/water emulsions. These emulsions can influence the calculated apparent viscosities (based on differential pressures) and may contribute to reducing CO_2 mobility.

4. Conclusions

This work presented a multi-scale investigation of hybrid nanoparticle–surfactant foam for CO_2 mobility control for CO_2 EOR and CO_2 storage. High-pressure micromodel experiments and high-pressure/high-temperature core floods evaluated a hybrid surfactant and nanoparticle foaming solution and foaming solutions with only surfactant or nanoparticles, in the presence and absence of oil. The following conclusions can be drawn:

- Direct pore-scale observations of dense phase CO_2 injection into a micromodel saturated with foaming solutions containing only surfactant or a hybrid nanoparticle–surfactant foaming solution revealed snap-off as the primary foam generation mechanism and leave-behind as a secondary foam generation mechanism.
- At the pore-scale, foam readily generated in areas where oil was not present and oil/water emulsions initially occupied pores without foam present.
- All foaming solutions containing surfactant generated foam in the presence and absence of oil, whereas foaming solution only containing nanoparticles did not. Thus, surfactant was the main foam generator and nanoparticles may be more important for foam stabilization.
- Foam strength was not sensitive to nanoparticle concentration when used together with surfactant in the tested foaming solutions.
- At the core-scale, all foaming solutions rapidly generated foam in the presence of residual oil.
- Foam apparent viscosity values with the hybrid foaming solution, in the presence of oil, were nearly three times as high as the experiments without oil. This was related to the development of oil/water emulsions, which were likely stabilized by the foaming agents.
- A link is proposed between direct pore-scale visual observations and quantitative core-scale measurements. The combined influence of stable oil/water emulsions and foam (CO_2/water emulsions) may be beneficial for increasing the capillary number by achieving higher apparent viscosity and lower interfacial tension.

- The experiments in this work were characterized by a period of rapid foam generation during drainage-like CO_2 injection and a period of foam coalescence during prolonged CO_2 injection. The decline in foam strength is related to the development of open CO_2 flow paths through the generated foam.
- Increased apparent viscosities with foam reduced CO_2 mobility at multiple length scales, which can improve volumetric sweep efficiency in field-scale CO_2 EOR and CO_2 storage processes.

Author Contributions: Conceptualization, Z.P.A., T.F., J.G., M.F.; methodology, Z.P.A., T.F., J.G., B.B., M.F.; formal analysis, Z.P.A., T.F., J.G.; investigation, T.F., B.B., A.S.; writing—original draft preparation, Z.P.A.; review and editing, T.F., J.G., B.B., M.F.; supervision, Z.P.A., J.G., M.F.; project administration, M.F.; funding acquisition, M.F. All authors have read and agreed to the published version of the manuscript.

Funding: This research was funded by the Norwegian Research Council, grant numbers 268216, 294886 and 301201.

Conflicts of Interest: The authors declare no conflict of interest.

References

1. IPCC. *Climate Change 2014: Chapter 1: Mitigation of Climate Change. Contribution of Working Group III to the Fifth Assessment Report of the Intergovernmental Panel on Climate Change*; Victor, D.G., Zhou, D., Ahmed, E.H.M., Dadhich, P.K., Olivier, J.G.J., Rogner, H.-H., Sheikho, K., Yamaguchi, M., Eds.; Cambridge University Press: Cambridge, UK; New York, NY, USA, 2014.
2. Enick, R.M.; Olsen, D.K.; Ammer, J.R.; Schuller, W. *Mobility and Conformance Control for CO2 EOR via Thickeners, Foams, and Gels—A Detailed Literature Review of 40 Years of Research*; DOE/NETL-2012/1540; USDOE, National Energy Technology Laboratory: Washington, DC, USA, 2012.
3. Hoefner, M.L.; Evans, E.M. CO_2 Foam: Results from Four Developmental Field Trials. *SPE Reserv. Eng.* **1995**, *10*, 273–281. [CrossRef]
4. Martin, F.D.; Stevens, J.E.; Harpole, K.J. CO_2-Foam Field Test at the East Vacuum Grayburg/San Andres Unit. *SPE Reserv. Eng.* **1995**, *10*, 266–272. [CrossRef]
5. Rossen, W.R. Foams in Enhanced Oil Recovery. In *Foams Theory, Measurements, and Applications*; Prud'homme, R.K., Khan, S.A., Eds.; Marcel Dekker, Inc.: New York, NY, USA, 1996; Volume 57, Chapter 11, pp. 414–464.
6. David, A.; Marsden, S.S. The Rheology of Foam. In Proceedings of the 44th Annual Fall Meeting of SPE of AIME, Denver, CO, USA, 28 September–1 October 1969. [CrossRef]
7. Espinosa, D.; Caldelas, F.; Johnston, K.; Bryant, S.L.; Huh, C. Nanoparticle-Stabilized Supercritical CO_2 Foams for Potential Mobility Control Applications. In Proceedings of the SPE Improved Oil Recovery Symposium, Tulsa, OK, USA, 24–28 April 2010.
8. Rognmo, A.U.; Al-Khayyat, N.; Heldal, S.; Vikingstad, I.; Eide, Ø.; Fredriksen, S.B.; Alcorn, Z.P.; Graue, A.; Bryant, S.L.; Kovscek, A.R.; et al. Performance of Silica Nanoparticles in CO_2-Foam for EOR and CCUS at Tough Reservoir Conditions. In Proceedings of the SPE Norway One-Day Seminar, Bergen, Norway, 18 April 2019.
9. Ogolo, N.A.; Olafuyi, O.A.; Onyekonwu, M.O. Enhanced Oil Recovery Using Nanoparticles. In Proceedings of the SPE Saudi Arabia Section Technical Symposium and Exhibition, Al-Khobar, Saudi Arabia, 8–11 April 2012.
10. Yu, J.; Wang, S.; Liu, N.; Lee, R. Study of Particle Structure and Hydrophobicity Effects on the Flow Behavior of Nanoparticle-Stabilized CO_2 Foam in Porous Media. In Proceedings of the SPE Improved Oil Recovery Symposium, Tulsa, OK, USA, 12–16 April 2014.
11. Bennetzen, M.; Mogensen, K. Novel Applications of Nanoparticles for Future Enhanced Oil Recovery. In Proceedings of the International Petroleum Technology Conference, Kuala Lumpur, Malaysia, 10–12 December 2014.
12. Ettinger, R.A.; Radke, C.J. Influence of Texture on Steady Foam Flow in Berea Sandstone. *SPE Reserv. Eng.* **1992**, *7*, 83–90. [CrossRef]
13. Chen, M.; Yortos, Y.C.; Rossen, W.R. A Pore-Network Study of the Mechanisms of Foam Generation. In Proceedings of the SPE Annual Technical Conference and Exhibition, Houston, TX, USA, 26–29 September 2004. SPE-90939.
14. Farajzadeh, R.; Andrianov, A.; Krastev, R.; Hirasaki, G.J.; Rossen, W.R. Foam–oil interaction in porous media: Implications for foam assisted enhanced oil recovery. *Adv. Colloid Interface Sci.* **2012**, *183*, 1–13. [CrossRef] [PubMed]

15. Harkins, W.D.; Feldman, A. Films: The spreading of liquids and the spreading coefficient. *J. Am. Chem. Soc.* **1922**, *44*, 2665–2685. [CrossRef]

16. Schramm, L.L.; Novosad, J.J. Micro-visualization of Foam Interactions with Crude Oil. *Colloids Surf.* **1990**, *46*, 21–43. [CrossRef]

17. Marsden, S.S. *Foams in Porous Media—SUPRI TR-49*; U.S. Department of Energy Topical Report; 1986. Available online: https://www.osti.gov/biblio/5866567 (accessed on 27 June 2020).

18. Talebian, S.H.; Masoudi, R.; Tan, I.M.; Zitha, P.L.J. Foam Assisted CO_2 EOR; Concepts, Challenges and Applications. In Proceedings of the SPE Enhanced Oil Recovery Conference, Kuala Lumpur, Malaysia, 2–4 July 2013.

19. Farajzadeh, R. Enhanced Transport Phenomena in CO_2 Sequestration and CO_2 EOR. Ph.D. Thesis, Faculty of Civil Engineering and Geosciences, Delft University of Technology, Delft, The Netherlands, 2009.

20. Kristiansand, T.S.; Holt, T. Properties of Flowing Foam in Porous media Containing Oil. In Proceedings of the SPE/DOE Improved Oil Recovery Symposium, Tulsa, OK, USA, 22–24 April 1992.

21. Kovscek, A.R.; Radke, C.J. Fundamentals of Foam Transport in Porous Media. In *Foams: Fundamentals and Applications in the Petroleum Industry*; American Chemical Society: Washington, DC, USA, 1994; pp. 115–163.

22. Saint-Jalmes, A. Physical chemistry in foam drainage and coarsening. *Soft Matter* **2006**, *2*, 836–849. [CrossRef]

23. Marchalot, J.; Lambert, J.; Cantat, I.; Tabeling, P.; Jullien, M.-C. 2D foam coarsening in a microfluidic system. *EPL (Europhys. Lett.)* **2008**, *83*, 64006. [CrossRef]

24. Falls, A.; Musters, J.; Ratulowski, J. The Apparent Viscosity of Foams in Homogeneous Bead Packs. *SPE Reserv. Eng.* **1989**, *4*, 155–164. [CrossRef]

25. Farajzadeh, R.; Lotfollahi, M.; Eftekhari, A.A.; Rossen, W.R.; Hirasaki, G.J.H. Effect of Permeability on Implicit-Texture Foam Model Parameters and the Limiting Capillary Pressure. *Energy Fuels* **2015**, *29*, 3011–3018. [CrossRef]

26. Buchgraber, M.; Al-Dossary, M.; Ross, C.M.; Kovscek, A.R. Creation of a dual-porosity micromodel for pore-level visualization of multiphase flow. *J. Pet. Sci. Eng.* **2012**, *86*, 27–38. [CrossRef]

27. Hornbrook, J.W.; Castanier, L.M.; Pettit, P.A. Observation of Foam/Oil Interactions in a New, High-Resolution Micromodel. In Proceedings of the SPE Annual Technical Conference and Exhibition, Dallas, TX, USA, 6–9 October 1991; Society of Petroleum Engineers: Houston, TX, USA, 1991.

28. Benali, B. Quantitative Pore-Scale Analysis of CO_2 Foam for CCUS. Master's Thesis, University of Bergen, Bergen, Norway, December 2019. Available online: http://hdl.handle.net/1956/21300 (accessed on 27 June 2020).

29. Bradski, G. The OpenCV Library. *Dr. Dobb's J. Softw. Tools* **2000**, *25*, 120–125.

30. Føyen, T.; Alcorn, Z.P.; Fernø, M.A.; Holt, T. 2020 CO_2 Mobility Reduction Using Foam Stabilized by CO_2- and Water-Soluble Surfactants. *J. Pet. Sci. Eng.* **2020**. [CrossRef]

31. Hirasaki, G.J.; Lawson, J.B. Mechanisms of Foam Flow in Porous Media: Apparent Viscosity in Smooth Capillaries. *SPE J.* **1985**, *25*, 176–190. [CrossRef]

32. Jones, S.A.; Laskaris, G.; Vincent-Bonnieu, S.; Farajzadeh, R. Surfactant Effect on Foam: From Core Flood Experiments to Implicit-Texture Foam- Model Parameters. In Proceedings of the SPE Improved Oil Recovery Conference, Tulsa, OK, USA, 11–13 April 2016; SPE-179637. [CrossRef]

33. Føyen, T.; Brattekås, B.; Fernø, M.; Barrabino, A.; Holt, T. Increased CO_2 storage capacity using CO_2-foam. *Int. J. Greenh. Gas Control* **2020**, *96*, 103016. [CrossRef]

34. Chambers, K.T.; Radke, C.J. *Interfacial Phenomena is Oil Recovery*; Morrow, N.R., Ed.; Marcel Dekker: New York, NY, USA, 1990.

35. Ranshoff, T.C.; Radke, C.J. Mechanics of Foam Generation in Glass Bead Packs. *SPE Reserv. Eng.* **1988**, *3*, 573–585. [CrossRef]

36. Rossen, W.R. Theory of mobilization pressure gradient of flowing foams in porous media: III. Asymmetric lamella shapes. *J. Colloids Interface Sci.* **1990**, *136*, 38–53. [CrossRef]

37. Falls, A.H.; Hirasaki, G.J.; Patzek, T.W.; Gauglitz, D.A.; Millerand, D.D.; Ratulowski, T. Development of a Mechanistic Foam Simulator: The Population Balance and Generation by Snap-Off. *SPE Reserv. Eng.* **1988**, *3*, 884–892. [CrossRef]

38. Kovscek, A.R.; Patzek, T.W.; Radke, C.J. A Mechanistic Population Balance Model for Transient and Steady-State Foam Flow in Boise Sandstone. *Chem. Eng. Sci.* **1995**, *50*, 3783–3799. [CrossRef]

39. Simjoo, M.; Zitha, P.L.J. Modeling and Experimental Validation of Rheological Transition during Foam Flow in Porous Media. *Transp. Porous Media* **2020**, *131*, 315–332. [CrossRef]

40. Ma, K.; Ren, G.; Mateen, K.; Moreland, D.; Cordelier, P. Modeling techniques for foam flow through porous media. *SPE J.* **2015**, *20*, 453–470. [CrossRef]

41. Gauteplass, J.; Follesø, H.N.; Graue, A.; Kovscek, A.R.; Fernø, M.A. Visualization of pore-level displacement mechanisms during CO_2 injection and EOR processes. In Proceedings of the EAGE IOR—17th European Symposium on Improved Oil Recovery, St. Petersburg, Russia, 16–18 April 2013.

42. Simjoo, M.; Dong, Y.; Andrianov, A.; Talanana, M.; Zitha, P.L.J. A CT Scan Study of Immiscible Foam Flow in Porous Media for EOR. In Proceedings of the SPE EOR Conference at Oil and Gas West Asia, Muscat, Oman, 16–18 April 2012.

 *nanomaterials*

Article

Experimental Investigation of Stability of Silica Nanoparticles at Reservoir Conditions for Enhanced Oil-Recovery Applications

Shidong Li [1,*], Yeap Hung Ng [1], Hon Chung Lau [1,2], Ole Torsæter [3,4]  **and Ludger P. Stubbs [1]**

[1] Institute of Chemical and Engineering Sciences, Agency for Science, Technology and Research (A*STAR), Singapore 627833, Singapore; ng_yeap_hung@ices.a-star.edu.sg (Y.H.N.); ceelhc@nus.edu.sg (H.C.L.); ludger_paul@ices.a-star.edu.sg (L.P.S.)

[2] Department of Civil and Environmental Engineering, National University of Singapore, Singapore 117576, Singapore

[3] PoreLab, Norwegian Center of Excellence, Norwegian University of Science and Technology (NTNU), 7031 Trondheim, Norway; ole.torsater@ntnu.no

[4] Department of Geoscience and Petroleum, Norwegian University of Science and Technology (NTNU), 7031 Trondheim, Norway

* Correspondence: li_shidong@ices.a-star.edu.sg

Received: 15 July 2020; Accepted: 29 July 2020; Published: 4 August 2020

Abstract: To be effective enhanced oil-recovery (EOR) agents, nanoparticles must be stable and be transported through a reservoir. However, the stability of a nanoparticle suspension at reservoir salinity and temperature is still a challenge and how it is affected by reservoir rocks and crude oils is not well understood. In this work, for the first time, the effect of several nanoparticle treatment approaches on the stability of silica nanoparticles at reservoir conditions (in the presence of reservoir rock and crude oil) was investigated for EOR applications. The stability of nanoparticle suspensions was screened in test tubes at 70 °C and 3.8 wt. % NaCl in the presence of reservoir rock and crude oil. Fumed silica nanoparticles in suspension with hydrochloric acid (HCl), polymer-modified fumed nanoparticles and amide-functionalized silica colloidal nanoparticles were studied. The size and pH of nanoparticle suspension in contact with rock samples were measured to determine the mechanism for stabilization or destabilization of nanoparticles. A turbidity scanner was used to quantify the stability of the nanoparticle suspension. Results showed that both HCl and polymer surface modification can improve nanoparticle stability under synthetic seawater salinity and 70 °C. Suspensions of polymer-modified nanoparticles were stable for months. It was found that pH is a key parameter influencing nanoparticle stability. Rock samples containing carbonate minerals destabilized unmodified nanoparticles. Crude oil had limited effect on nanoparticle stability. Some components of crude oil migrated into the aqueous phase consisting of amide-functionalized silica colloidal nanoparticles suspension. Nanoparticles modification or/and stabilizer are necessary for nanoparticle EOR application.

Keywords: nanoparticle stability; reservoir condition; reservoir rock; crude oil; nanoparticle agglomeration

---

1. Introduction

In recent years, nanotechnology research on enhanced oil recovery (EOR) has shown promising results in the laboratory. Some EOR experiments with silica nanoparticles have been performed and showed positive results in increasing oil recovery [1–4]. The proposed EOR mechanisms for silica nanoparticles include interfacial tension reduction, wettability alteration, plugging of pore channels, disjoining pressure and emulsification [5–8].

The prerequisite of these mechanisms working well in the reservoir is that nanoparticles are stable at reservoir conditions so that they can maintain their surface activities for EOR. The nanofluid preparation method is crucial for its stability. For EOR applications, the most commonly used base fluid is brine. In the stationary state, the settling velocity of small spherical particles in a suspension follows Stokes law [9]:

$$V = \frac{2R^2}{9\mu}\left(\rho_p - \rho_L\right)\cdot g \tag{1}$$

where V is the particle's settling velocity; R is the spherical particle's radius; μ is the liquid medium viscosity; ρ_p and ρ_L are the particle and the liquid medium density, respectively, and g is the acceleration of gravity. This equation explains the effect of gravity, buoyancy and viscous drag on the behavior of the suspended particles in the base fluid. According to Stokes law, the nanoparticle settling velocity decreases with: (1) reducing R, (2) increasing μ, the base fluid viscosity and (3) lessening the density difference between nanoparticles and base fluid [10].

There are two different techniques used to produce nanofluids. First, a two-step technique in which a dry nanoparticle powder is first produced and then dispersed in a base fluid. However, due to the high surface energy of nanoparticles, aggregation and clustering easily take place after preparation. Therefore, additional techniques are needed to minimize this problem, such as a high shear homogenization and ultrasound. Alternatively, a single-step technique can be applied wherein nanoparticles synthesis and nanofluid preparation are undertaken simultaneously. In the single-step method, there is no drying, storage, transportation, and dispersion of nanoparticles, so the aggregation of nanoparticles is minimized and the stability of the nanofluid can be improved [10].

However, due to van der Waals attraction between nanoparticles and high-temperature and high-salinity conditions in a reservoir, the injected nanoparticles may still undergo aggregation and agglomeration, thus losing their colloidal stability. In order to maintain long-term stability of a nanofluid in a reservoir, some physical or/and chemical treatments are necessary [10]. Generally, nanoparticles have to meet two principles to achieve good stability. The first is the diffusion principle: The nanoparticles are scattered by and dispersed into a liquid medium. The second is the Zeta potential principle: the absolute zeta potential value of a nanofluid must be larger than a specified value and have a sufficient repulsive force between the nanoparticles [11]. The commonly used techniques to increase nanofluid stability are summarized as follows. The first technique is pH control. The stability of a nanofluid is directly related to its electrokinetic properties. Strong repulsion forces can create a stable nanofluid through a high surface charge density [12–14]. Sofla et al. proposed "H$^+$ protection" theory. They showed that adding hydrochloric acid (HCl) into a nanofluid can effectively stabilize silica nanoparticles in seawater [15]. The second technique is addition of a surfactant. This is a general method used to improve the stability of nanoparticles in aqueous phase. The mechanism is that the hydrophobic surface of nanoparticles is covered with surfactant and changed to hydrophilic, thus the repulsion force will become larger and absolute value of zeta potential will increase [16,17]. Surfactant selection (cationic, anionic or non-ionic) is very important [18]. The third technique is surface modification (steric stabilization). This method utilizes the adsorption of large molecules like surfactants and polymers on the surface of nanoparticles to prevent aggregation. The adsorbed molecules lead to an increase of osmotic repulsion, resulting in higher colloidal stability of the nanofluid [19]. Ranka et al. achieved a stable nanofluid by modifying silica nanoparticles with zwitterionic polymers [20]. The fourth method is ultrasonication to break down the agglomeration and clustering of nanoparticles [21]. Therefore, particle size will be reduced remarkably.

After a nanofluid is prepared by using the aforementioned techniques, specific analytical methods are needed to evaluate the relative stability of the nanofluid samples:

(1) Optical visualization. This is a simple method to determine the stability of nanofluids in test tubes by optical inspection for sedimentation after certain periods of time [10]. This method is only applicable for low viscosity and transparent base fluids.

(2) Zeta potential measurement. This method is one of the most critical tests to validate the stability of nanofluids. If the zeta potential has a high absolute value, electrostatic forces repel nanoparticles from each other [22]. Generally, a nanofluid with a measured absolute zeta-potential value above 30 mV is regarded as stable [23]. However, zeta potential results may not be reliable for high salinity base fluids.

(3) Light transmission and scattering method. The intensity of transmitted and scattered light for a single particle is related to the particle volume. Particle size and sedimentation thickness can be measured in real time [8].

(4) Ultraviolet–visible (UV–Vis) spectrophotometry. The absorption of light by nanoparticles is used to calculate the concentration of nanoparticles and quantitatively evaluate the stability of the nanofluid [10].

(5) Scanning electron microscopy (SEM) and transmission electron microscopy (TEM). SEM and TEM are powerful tools to characteristic the shape, size and distribution of nanoparticles. In addition, cryogenic electron microscopy (Cryo-SEM and Cryo-TEM) can observe the real state of nanoparticles in the nanofluid, while SEM and TEM only work for dry samples. [10] Therefore, the aggregation nanoparticles in a nanofluid can be observed directly.

(6) Sedimentation balance method. In this method the tray of a sedimentation balance is immersed in the fresh nanofluid. The weight of nanoparticles sedimentation during a certain period of time is measured, hence the total nanoparticles sedimentation can be calculated accordingly [22].

(7) Three omega method. In this method the stability of the nanofluid can be evaluated by detecting the thermal conductivity growth caused by the nanoparticle sedimentation. Several stability tests refer to this method in the literature. [24–26]

For the application of nanofluids in EOR, high temperature, high salinity and the presence of divalent cations such as Mg^{2+}, Ca^{2+}, and Ba^{2+} present challenges to the colloidal stability. Several literature reports have studied the stability of nanoparticles at reservoir conditions for EOR applications. For example, Metin et al. observed a critical salt concentration for a given salt solution, below which the nanoparticles have a good stability and above which aggregation of nanoparticles occur and sedimentation is observed at the bottom of the samples. They also reported that divalent cations played a more important role in destabilizing nanoparticles than monovalent cations, and that elevated temperatures accelerate the nanoparticle aggregation process [27]. Sofla et al. studied silica nanoparticle stability in synthetic seawater and observed that HCl can effectively stabilize nanoparticles in seawater [15]. Ranka et al. functionalized silica nanoparticles with zwitterionic polymers to undergo a structural transition from a collapsed globule to a more open coil-like structure with increasing ionic strength and temperature. The surface functionalized silica nanoparticles exhibited long-term stability at salinities up to 120 g/L at 90 °C [20]. However, in the aforementioned studies only the effect of salts and temperature on the stability of the nanoparticles was considered. Two other important factors during an EOR process, namely, the effect of reservoir rocks and crude oils, may also play a role in the stability of nanoparticles. In this study several reservoir rocks and crude oil samples from different oil fields were used to investigate their effects on nanoparticle stability under high-temperature and high-salinity conditions. Nanofluid EOR has shown its potential of increased oil recovery in the laboratory by using both coreflood and microfluidic flooding experiments. Hendraningrant et al. has reported that fumed silica nanoparticles could increase oil recovery by 2% to 10% with 20 coreflood experiments [28]. Khezrnejad et al. performed microfluidic flooding experiments with silica nanoparticles and found nanofluid flooding increased oil recovery by more than 10% compared to water flooding [29]. However, these two experiments were conducted under room temperature and used NaCl brine and deionized water, which is favorable for the stability of nanoparticles. This study also tried to find a solution to stabilize silica nanoparticles under reservoir conditions, making silica nanoparticles also have a potential to increase oil recovery for a field application.

2. Experimental Materials

2.1. Nanoparticle

Fumed hydrophilic silica nanoparticles (FNP) were provided by Evonik Industries (Germany). The primary particle size of FNP is about 7 nm, but in suspension nanoparticles will aggregate and the size may increase to more than 100 nm. The specific surface area of FNP is around 300 m^2/g. Amide-functionalized colloidal silica nanoparticles (ANP) dispersion was purchased from Sigma-Aldrich (Singapore). The particle size of ANP is less than 30 nm. Surface-modified FNP with zwitterionic polymer (FNP-MD) were prepared according to the procedure described in Section 3.4. The nanoparticles were characterized by Cryogenic Transmission Electron Microscopy (Cryo-TEM) (FEI Titan Krios, Thermo Scientific™, Waltham, MA, USA) and Scanning Electron Microscope (SEM) (JSM-6700F, JEOL, Japan). Images are shown in Figures 1 and 2. For Cryo-TEM imaging, the sample (5 μL) was applied onto a grid (Quantifoil, R2/2, Holey carbon film; freshly glow-discharged prior to use at 20 mA for 60 s) without dilution. Excess of sample was blotted away with filter paper to leave a thin film on the grid before being vitrified in liquid ethane. Cryo-TEM measurements were performed on an FEI Titan Krios equipped with automated sample loader and a field-emission gun (FEG) operating at 300 kV. Images were recorded with Falcon II camera (4 × 4) with magnification of 29,000 and pixel size of 2.873 Å [30]. For SEM imaging, diluted nanofluid was placed on a coverslip. After dying, the sample was sputtered with gold and ready for SEM imaging. The high voltage was 5.0 kV, working distance ranged from 7.4 mm to 9.1 mm and secondary electron image was used.

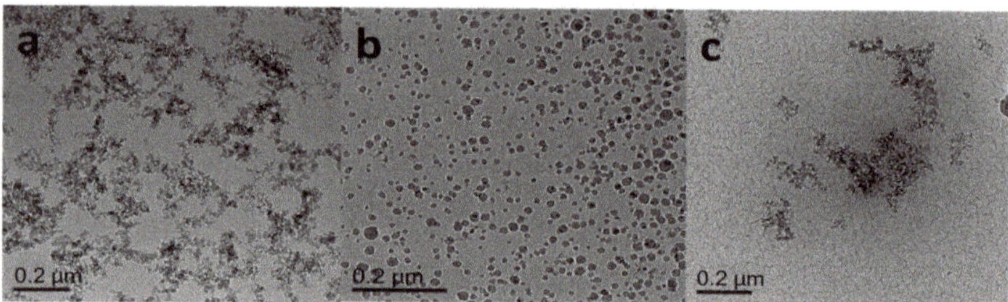

Figure 1. Cryo-transmission electron microscope (TEM) image of nanoparticles: (**a**) fumed hydrophilic silica nanoparticles (FNP); (**b**) amide-functionalized colloidal silica nanoparticles (ANP); (**c**) surface-modified FNP with zwitterionic polymer (FNP-MD).

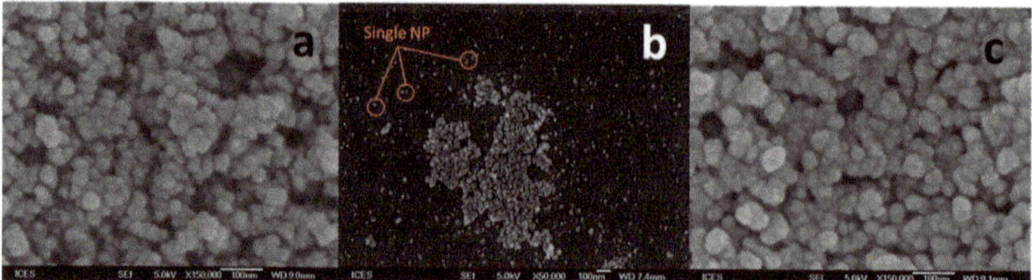

Figure 2. Scanning electron microscope (SEM) image of nanoparticles: (**a**) FNP; (**b**) ANP; (**c**) FNP-MD.

2.2. Nanoparticles Suspensions

Five concentrations (0.1, 0.2, 0.3, 0.4 and 0.5 wt. %) of nanoparticles dispersed in lab made synthetic seawater (SSW) of 3.8 wt. % salinity were used. The recipe of SSW is given in Table 1.

For FNP and FNP-MD, nanoparticles were weighed and then dispersed in SSW by using a sonicator (40W for 20 min). For HCl-stabilized FNP suspension samples (FNP-HCl), concentrated aqueous HCl was added to the suspension of FNP until pH 2.0.

Table 1. Synthetic seawater recipe.

Salts	Concentration (g/L)	Salts	Concentration (g/L)
$CaCl_2 \cdot 2H_2O$	1.76	Na_2SO_4	4.81
$MgCl_2 \cdot 6H_2O$	11.23	NaCl	27.03

The ANP suspensions were prepared by a dilution of concentrated ANP dispersion sample. The size of nanoparticles such in suspension was measured using ZetaSizer Nano dynamic light scattering (DLS) (Malvern Panalytical, Malvern, UK) in a standard 12 mm polystyrene cuvette. The samples was measured at 25 °C without any dilution at 173° backscatter [30]. The measured particle size for FNP, FNP-MD and ANP are 142.5 nm, 168.3 nm and 21.4 nm, respectively.

2.3. Reservoir Rocks

Two Berea sandstones (BSS) (Kocurek Industries, Inc, Caldwell, TX, US) with low (BSS1) and high (BSS2) permeability were used. Chalk, limestone and shale were also used in this study. The mineral composition (by X-ray diffraction) for these rock samples are shown in Table 2. Pure quartz sands (Sigma-Aldrich, Singapore) were used as well for control experiments.

Table 2. Mineral components of rock samples.

%	BSS1	BSS2	Chalk	Limestone	Shale
SiO_2 Quartz	90.87	90.31	–	0.69	2.84
Na $AlSi_3O_8$ Albite	0.89	1.89	–	–	–
$KAlSiO_8$ Sanidine	0.46	0.74	–	–	–
$KAL_2(Si_3Al)O_{10}(OH,F)_2$ Muscovite	3.75	3.6	–	–	2.07
$Al_2SiO_5(OH)_4$ Kaolinite	1.92	2.29	–	–	1.48
$(Mg,Fe)_5Al(Si_3Al)O_{10}(OH)_8$ Clinochlore	0.97	0.89	–	–	–
$CaMg(CO_3)_2$ Dolomite	1.13	0.28	–	–	1.89
$CaCO_3$ Calcite	–	–	100	99.31	89.28
$Ca_5(PO4)_3(OH)_2$ Apatite	–	–	–	–	2.44

2.4. Oils

Decane and eight different types of crude oils (CO1-7) were used in this experiment. Crude oil properties were measured and are given in Table 3.

Table 3. Crude oil properties.

	CO1	CO2	CO3	CO4	CO5	CO6	CO7
Saturates	60.66	75.72	84.61	38.48	74.79	26.37	32.94
Aromatics	10.52	17.43	12.76	56.13	19.79	51.90	52.26
Resins	9.68	3.85	2.32	5.33	5.06	20.76	14.80
Asphaltenes	19.15	3.00	0.31	0.07	0.36	1.09	0.00
Density@70 °C (kg/m^3)	826	793	786	852	782	926	902
Viscosity@70 °C (mPa·s)	7.66	2.47	1.90	2.43	1.51	35.61	13.06
Sulphur Content (ppm)	657	491	655	1300	574	3325	1900
Total Acid Number (mg/mg KOH)	0.064	0.475	0.103	0.260	0.239	1.720	1.702

3. Experimental Methods

3.1. Nanoparticles Suspension Stability

Nanoparticle suspension stability tests were performed by using visual observation and turbidity scanning. For each type of nanoparticle (FNP, FNP-HCl, FNP-MD and ANP), samples with five different concentrations (0.1, 0.2, 0.3, 0.4 and 0.5 wt. %) were prepared and put into test tubes, which were then placed into a heating cabinet at 70 °C. Photographs of all test tubes were taken after certain period of time to show the stability of samples over time. A Turbiscan LAB colloidal stability analyzer (Formulaction Inc., Toulouse, France) was used to quantify nanoparticle suspension stability. Samples of nanoparticle suspension were scanned at 60 °C with an 880 nm near-infrared light-emitting diode (LED) source and the transmission and backscattered signals were registered by detectors. Since nanoparticle size can affect these signals, delta transmission and backscattering were used to determine nanoparticle stability. The dimensionless turbidity scan index (TSI) defined by the Turbiscan software (TurbiSoft Lab, 2.2.0.82-2, Toulouse, France) was used to quantify nanoparticles suspension stability. Some samples need longer scanning time. They were scanned continuously for the first 10 days then were taken out of the instrument and put into the heating cabinet with a temperature of 60 °C. They were placed back for a single scan every second day. Calculation of *TSI* is based on comparing each scan to the previous one for the selected height and dividing the result by the total selected height in order to obtain a result which does not depend on the quantity of product in the measuring cell. The lower the *TSI* value, the better is the stability of the sample. The following equation was used to calculate the *TSI* [31].

$$TSI = \sum_i \frac{\sum_h \left| scan_i(h) - scan_{i-1}(h) \right|}{H} \tag{2}$$

3.2. Effect of Reservoir Rocks and Crude Oils on Nanoparticle Stability

FNP, FNP-MD and ANP suspensions at 0.1 wt. % concentration were used in this stability test. The nanoparticle suspension (7 mL) was put into a test tube and 0.2 gm of crushed rock sample was added. Pure quartz was used as a benchmark for the rock samples. Another set of samples was prepared with 5 mL nanoparticle suspension and 2 mL oil in test tubes. Decane was used as reference oil. The test tubes were placed inside a heated cabinet at 70 °C.

3.3. Particles Size and pH Measurement of Nanoparticles

The particle size distribution of the nanofluids was measured by DLS until nanoparticles were fully agglomerated. The pH of nanofluids was measured with a pH meter.

3.4. Preparation of Polymer Modified Nanoparticles

[2-(Methacryloyloxy)ethyl]dimethyl-(3-sulfopropyl) ammonium hydroxide (MEDSAH, 95%, Aldrich, Singapore), 2-(2-carboxyethylsulfanylthiocarbonylsulfanyl)propionic acid (CTA, 95%, Aldrich, Singapore), 3-(trimethoxysilyl) propyl methacrylate (MPS, 98%, Aldrich, Singapore), 2,2′-azobis(2-methylpropionamidine) dihydrochloride (V50, Wako Chemicals, Osaka, Japan), fumed silica (FNP), toluene (high-performance liquid chromatography (HPLC), VWR Chemicals) were used as received.

In a sealed Schlenk flask, fumed silica was dried under vacuum at 120 °C for at least 24 h before being used. As illustrated schematically in Figure 3, to a dispersion of dried silica (2 g) in toluene (100 mL) being stirred (1000 rpm) and under argon protection, MPS (3 g) was added using an argon-purged syringe. The mixture was then heated to 100 °C for 12 h. After cooling to room temperature, the dispersion was centrifuged and washed with fresh toluene (3 times), followed by ethanol rinsing (3 times). The methacrylate-functionalized silica (SiO$_2$-MPS) was dried under vacuum at 60 °C for 12 h. As-prepared SiO$_2$-MPS (0.5 g) was dispersed in deionized water (25 mL) under sonication for 2 h. MEDSAH (2 g), CTA (36.6 mg), V50 (7.8 mg) (MEDSAH/CTA/V50 = 50/1/0.2) were placed into a dry Schlenk flask. The flask was sealed with a rubber septum and subjected to 4 vacuum/argon cycles. SiO_2-MPS dispersion was deoxygenated by bubbling argon for about 1 h and added to the Schlenk flask using an argon-purged syringe. The mixture was heated at 60 °C under a stirring rate of 700 rpm. After 24 h, the dispersion was cooled to room temperature, and washed with deionized water for 8 times. The polymer-coated silica (SiO_2-MEDSAH) was dried under vacuum at 80 °C for 24 h.

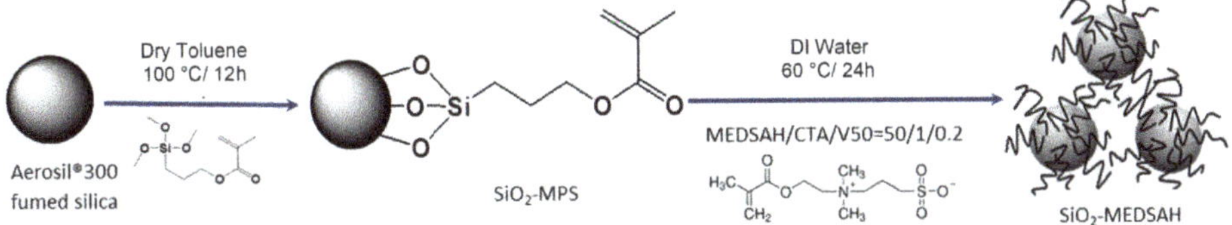

Figure 3. Schematic of preparation polymer modified nanoparticles.

4. Results and Discussion

4.1. Nanoparticle Suspension Stability Tests

Four types of nanoparticle suspensions (FNP, FNP-MD, ANP and FNP-HCl) were used in the stability tests. The stability of nanofluids was determined by visual observation and quantified by turbidity scanning analysis. The time when nanoparticle agglomeration occurred for each sample was recorded and is given in Table 4. It can be seen that the unmodified fumed silica (FNP) suspensions had the worst stability. All samples agglomerated and settled within one day. The polymer modified fumed silica (FNP-MD) suspensions had the best stability. All five samples with different concentrations were still stable after thirty days. Adding HCl to fumed silica (FNP-HCl) delayed nanoparticle agglomeration. For fumed silica suspensions, stability increased with decreasing nanoparticle concentration. However, ANP colloidal suspensions showed the opposite trend. The reason is that acid was used to stabilize concentrated ANP suspension by manufacturer, so that 0.5 wt. % diluted sample (pH 4.1) had a lower pH than the 0.1 wt. % diluted sample (pH 5.3).

Table 4. Nanoparticles stability screening test.

NPs Conc. (wt. %)	Days Until Agglomeration Was Observed			
	FNP	FNP-MD	FNP-HCl	ANP
0.1	1	–	–	1
0.2	1	–	25	8
0.3	<1	–	15	25
0.4	<1	–	12	–
0.5	<1	–	10	–

–: no nanoparticles agglomeration was observed within 30 days.

The turbidity scanning tests were performed for nanoparticle suspensions of 0.1 wt. % concentration at 60 °C (instrument limit). As an example, Figure 4 shows the delta transmission (difference of transmission single strength between subsequent scans and initial scan) scanning results for FNP and FNP-MD samples, in which FNP was scanned for one day and FNP-MD was scanned for 10 days continuously. In Figure 4, the blue curve indicates the initial scan and the red curve indicates the final scan. The X axis indicates the sample height (left = bottom of the vial) and the Y axis indicates the delta transmission single strength (in %). As shown in Figure 4a, the FNP nanofluid underwent a strong change in stability. Between 0 h to around 14 h (blue to green), the transmission single strength reduced slightly, which showed that nanoparticle agglomeration occurred and average particle size increased, but nanoparticles were still suspended. However, when the particle size increases to a critical value where gravity dominates, the sedimentation of nanoparticles occurs. In Figure 4a, this phenomenon happened after around 14:30 h and a marked change of delta transmission curves was observed. With more nanoparticle sedimentation, the transmission single strength declined strongly at the bottom of the FNP sample. On the contrary, due to clarification the transmission single strength in the middle and top of the sample increased back to the initial value. In Figure 4b, no significant change of transmission single strength was observed for polymer modified FNP nanofluid over ten days, indicating that it has a good stability compared with the unmodified FNP nanofluid. This result also showed that after surface modification the repulsion force was strong enough to keep nanoparticles apart from each other, so that no agglomeration was detected over ten days.

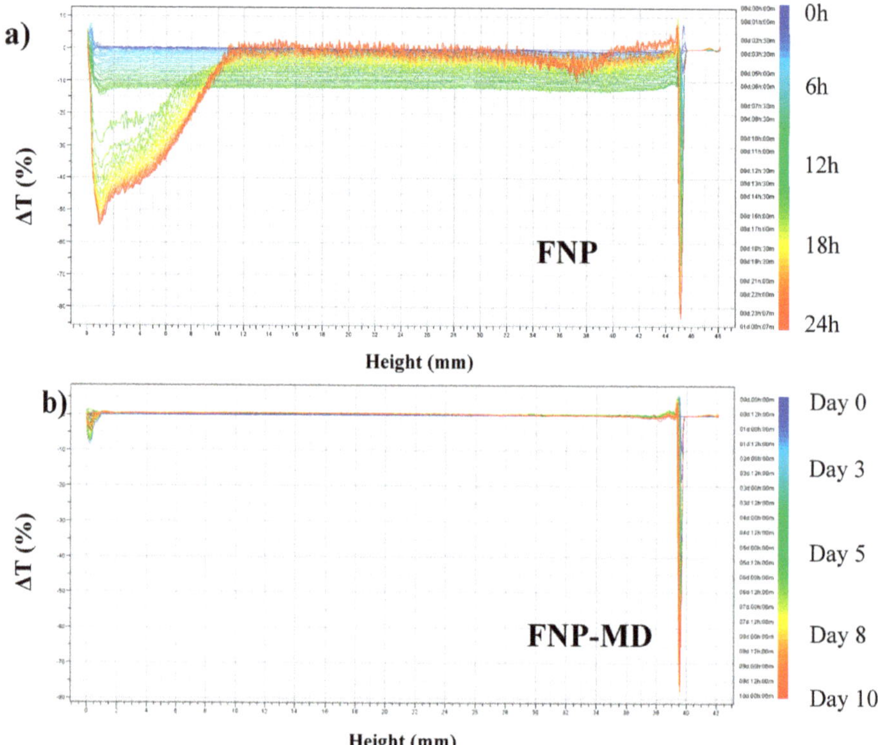

Figure 4. Delta transmission scanning results of nanofluid samples. (**a**): FNP (for one day); (**b**): FNP-MD (for 10 days).

The TSI values of the samples were calculated based on transmission scanning results and plotted versus time (Figure 5). It can be seen that both FNP-HCl and FNP-MD had relatively low TSI values over 30 days, which means that these two samples had a good stability. However, TSI values for FNP and ANP samples increased very fast to a plateau and stayed constant after one day, which indicated that the nanoparticles had agglomerated, settled and lost their stability. This result is consistent with test tube observations.

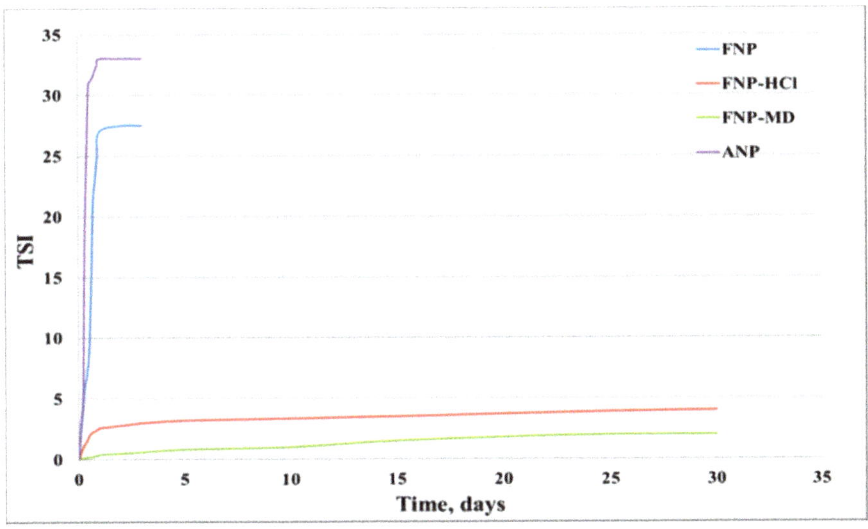

Figure 5. Turbidity scan index (TSI) results for nanoparticles suspensions.

4.2. Effect of Reservoir Rocks on Nanoparticle Stability

Five reservoir rocks (Berea sandstone 1 and 2, chalk, limestone and shale), as well as pure quartz sand as a control mineral, were used in this experiment. FNP, FNP-MD and ANP suspensions with 0.1 wt. % concentration were used. Photographs of all samples in test tubes were taken for nanoparticle stability observation. Particle size and pH of nanoparticle suspensions were measured to investigate the mechanisms for stabilization or destabilization of nanoparticles due to the presence of rock samples. The photographs for FNP, FNP-MD and ANP suspensions with rock samples are shown in Figures 6–8. The results of pH and particle size measurements are shown in Figures 9–11. As evident from Figures 6 and 7, reservoir rocks had a significant effect on the stability of FNP and ANP, especially for limestone, chalk and shale. These three rocks have very high calcite content (see Table 2), so they could react with H^+ in the nanoparticle suspensions. In Figures 9 and 10, the pH value of limestone, chalk and shale samples increased to pH 7 quickly, and this neutral pH is not favorable for nanoparticle stability. Thus, fast nanoparticle agglomeration was observed. The Berea sandstones also had an effect on FNP and ANP suspension stability. Nanoparticle agglomeration was observed for BSS1 earlier than for BSS2. The reason might be the higher dolomite content in BSS1 (Table 2), which leads to a quicker pH increase of the suspensions, therefore the nanoparticle size in BSS1 samples increased faster than in BSS2 samples (Figures 9 and 10). No significant effect of quartz sands on FNP stability was observed, while quartz sands stabilized ANP for longer time (Figure 7) compared with the stability result of ANP without quartz sands shown in Table 4; and particle size increased much slower than others (Figure 10). The reason for this phenomenon is unknown and needs further study. For FNP-MD suspensions (Figure 8), there was almost no effect of the rocks on the nanofluid stability, except for limestone. As shown in Figure 11, the pH increased in all reservoir rock samples, but FNP-MD nanofluids were still stable after 30 days in the heating cabinet, even though the particle size increased in chalk and shale samples. For some unknown reason, limestone destabilized the FNP-MD suspension. Since limestone has a similar mineral composition like chalk, more tests need to be done to determine the mechanism of this phenomenon.

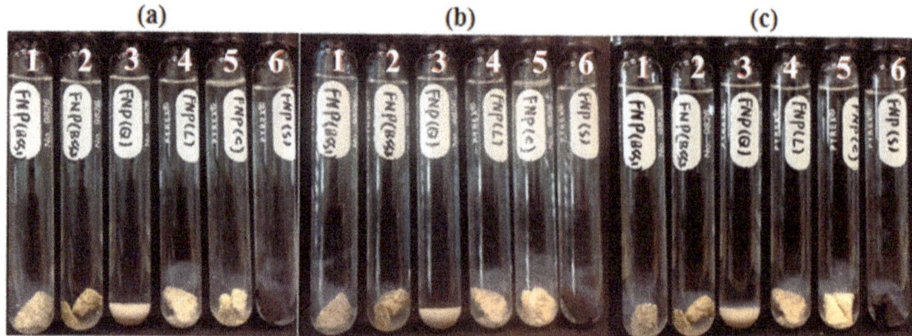

Figure 6. FNP suspensions with reservoir rocks: (**a**), Day 0; (**b**), Day 1; (**c**), Day 3. (1: BBS1; 2: BSS2; 3: quartz; 4: limestone; 5: chalk; 6: shale).

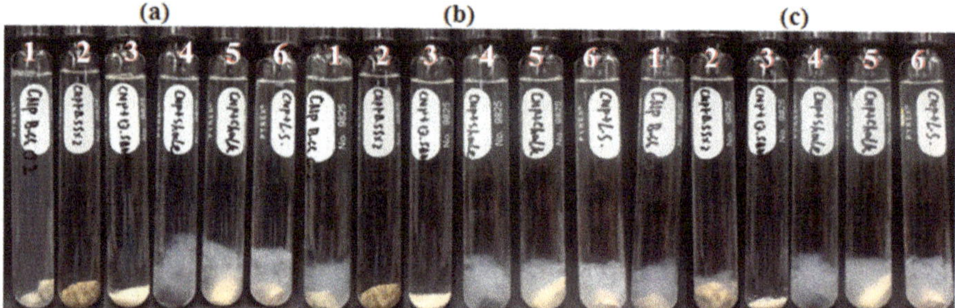

Figure 7. ANP suspensions with reservoir rocks: (**a**), Day 0; (**b**), Day 1; (**c**), Day 3. (1: BBS1; 2: BSS2; 3: quartz; 4: shale; 5: chalk; 6: limestone).

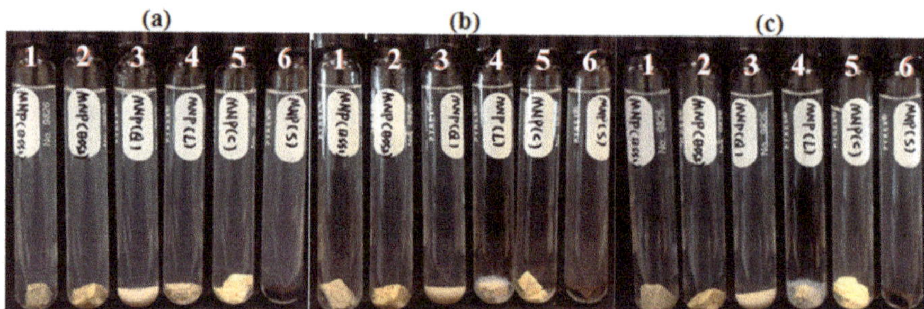

Figure 8. FNP-MD suspensions with reservoir rocks: (**a**), Day 0; (**b**), Day 7; (**c**), Day 30. (1: BBS1; 2: BSS2; 3: quartz; 4: limestone; 5: chalk; 6: shale).

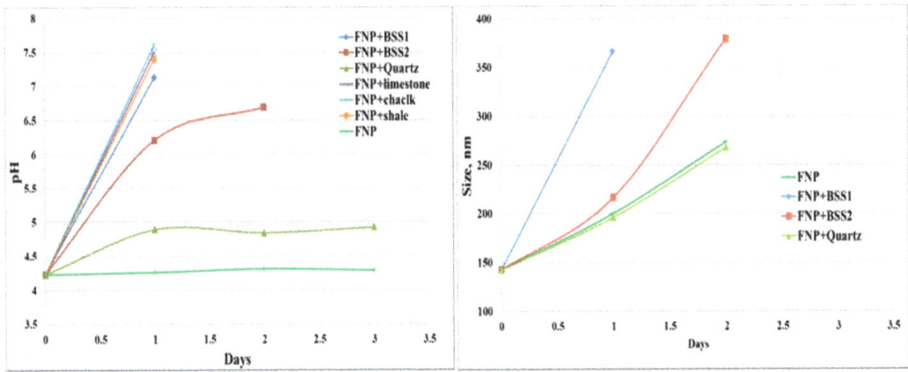

Figure 9. pH and particle size measurements for FNP suspension with different rocks.

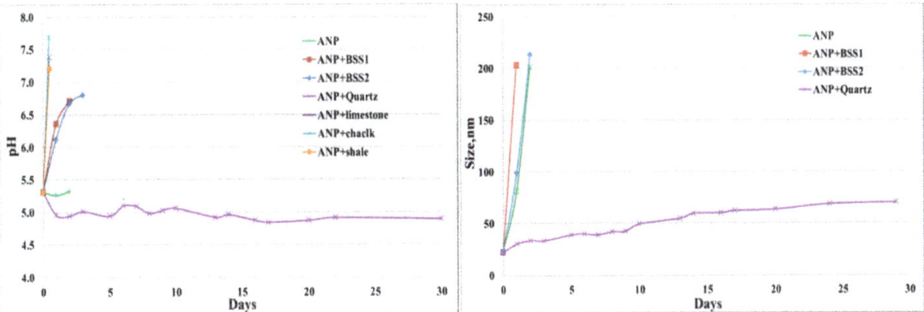

Figure 10. pH and particle size measurements for ANP suspension with different rocks.

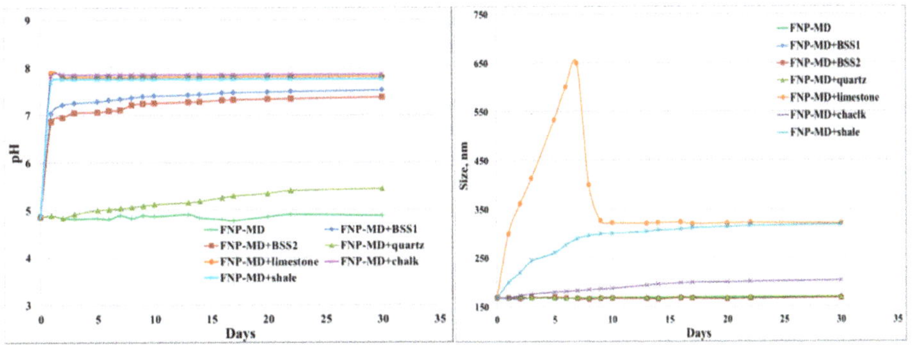

Figure 11. pH and particle size measurements for FNP-MD suspension with different rocks.

4.3. Effect of Crude Oils on Nanoparticle Stability

Seven crude oils with different properties were used in this study, decane was used as a control oil. FNP, FNP-MD and ANP suspensions with 0.1 wt. % concentration were used. Photographs of all samples in test tubes were taken for nanoparticle stability observation. For FNP nanofluids (Figure 12), crude oils had no obvious effect on their stability. Only nanoparticle agglomeration delay was observed in the CO4 sample on the first day. FNP-MD showed a good stability in the presence of crude oils and was still stable after 30 days in the heating cabinet (Figure 13). For ANP suspension, nanoparticle sedimentation occurred in CO1 and decane samples on day 1, and all samples aggregated on day 3 (Figure 14). For samples with CO2, CO3, CO4 and CO5, nanoparticle sedimentation appeared yellowish. This might be due to migration of some components of the crude oil into the aqueous suspension which adsorbed on the nanoparticle surface, which indicates that crude oil has the potential to affect the structure of nanoparticles in suspension.

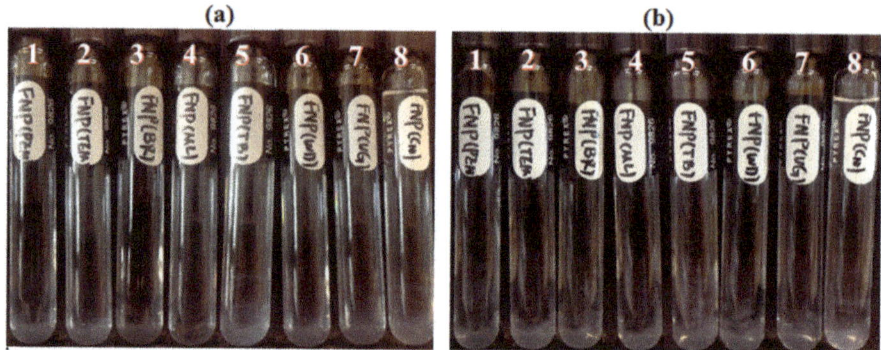

Figure 12. FNP suspensions stability screening with different type of oils: (**a**), Day 1; (**b**), Day 2; (1: CO1; 2: CO2; 3: CO3; 4: CO4; 5: CO5; 6: CO7; 7: CO6; 8: Decane).

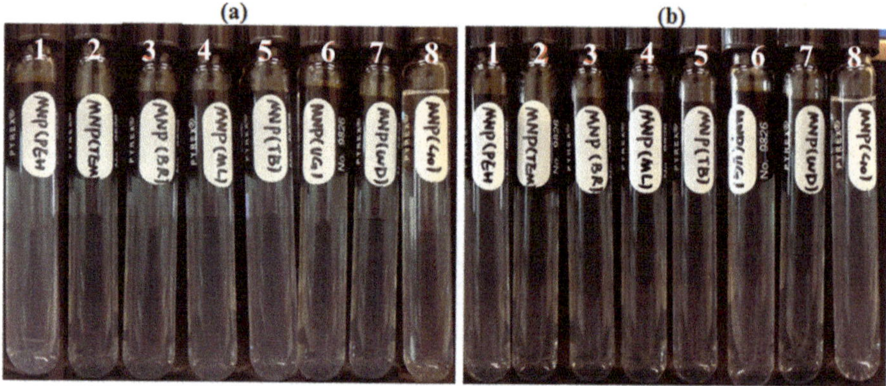

Figure 13. FNP-MD suspensions stability screening with different type of oils: (**a**), Day 1; (**b**), Day 30; (1: CO1; 2: CO2; 3: CO3; 4: CO4; 5: CO5; 6: CO6; 7: CO7; 8: Decane).

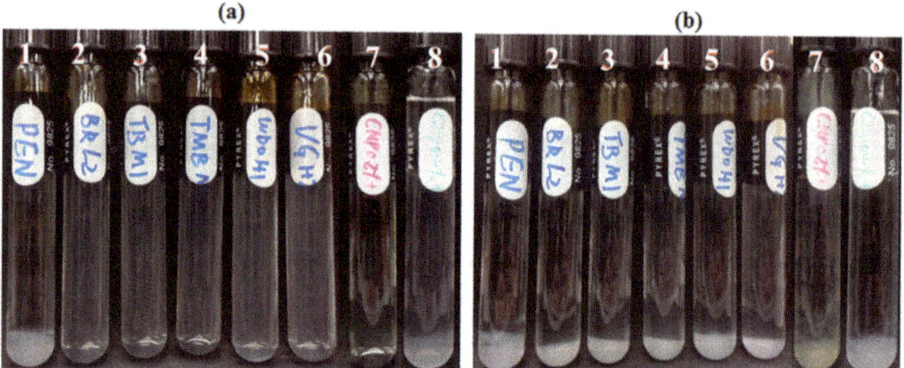

Figure 14. ANP suspensions stability screening with different type of oils: (**a**), Day 1; (**b**), Day 3; (1: CO1; 2: CO3; 3: CO5; 4: CO2; 5: CO7; 6: CO6; 7: CO4; 8: Decane).

5. Conclusions

Stability screening tests were performed for a range of unmodified and surface-modified silica nanoparticles, as well as silica nanoparticles with HCl as stabilizer by using methods of stability visualization and turbidity scanning. The results showed that surface modification with polymer and addition of HCl can improve stability of fumed silica nanoparticles under high-temperature and high-salinity conditions remarkably. The turbidity scanning method is a quite useful technique for nanoparticle stability studies, it can reveal nanoparticle agglomeration and sedimentation processes and also quantify nanoparticle stability. The effect of reservoir rock samples and crude oils on nanoparticle stability was also examined. It was found that chalk, limestone and shale destabilized fumed silica nanoparticles and colloidal silica nanoparticles quickly, while they had no obvious influence on the stability of polymer-modified fumed silica nanoparticles. It was also observed that the Berea sandstone containing more dolomite destabilized fumed silica nanoparticles and colloidal silica nanoparticles faster than low dolomite content Berea sandstone. The mechanism of these destabilizations is probably the reaction between carbonate minerals and H^+ of the nanofluid, leading to a pH increase that accelerated the agglomeration of nanoparticles. However, polymer-modified fumed silica nanoparticles showed a good long-term stability. Quartz sands could stabilize colloidal silica nanoparticle longer than the blank stability test without quartz sands in aqueous suspension. Crude oils had limited effect on fumed silica nanoparticle and colloidal silica nanoparticle stability and no obvious trend was observed, only agglomeration delay was observed in a few samples. Some components of crude oil migrated into the colloidal silica nanoparticle suspensions. Polymer-modified fumed silica nanoparticles had a good stability in the presence of crude oil over 30 days.

Author Contributions: Conceptualization, methodology, data analysis, investigation and validation, S.L.; sample nanoparticle modification, Y.H.N.; funding acquisition, project administration, L.P.S.; writing—original draft preparation, S.L. and Y.H.N.; writing—review and editing, L.P.S., H.C.L. and O.T. All authors have read and agreed to the published version of the manuscript.

Funding: This research was supported by the Agency for Science, Technology and Research (A*STAR), Singapore, under IAFPP Programme (project title: Advanced Functional Polymer Particle Technologies for the Oil and Gas Industry; grant no.: A18B4a0094).

Acknowledgments: The authors would like to thank the Agency for Science, Technology and Research (A*STAR), Singapore, for financial support as well as the Petroleum Engineering Professorship Grant from the Economic Development Board of Singapore. The authors thank Wendy Rusli for Cyro-TEM imaging, Andrew Lim Shin Boon for X-ray diffraction (XRD) measurement, Martin Karl Schreyer for XRD interpretation and Buana Girisuta for providing crude oil properties.

Conflicts of Interest: The authors declare no conflict of interest.

References and Note

1. Lau, H.C.; Yu, M.; Nguyen, U.P. Nanotechnology for oilfield applications: Challenges and opportunities. *J. Petrol. Sci. Eng.* **2017**, *157*, 1160–1169. [CrossRef]
2. Kong, X.; Ohadi, M.M. Applications of micro and nano technologies in the oil and gas industry—Overview of the recent progress. In Proceedings of the Abu Dhabi International Petroleum Exhibition and Conference, Abu Dhabi, UAE, 1–4 November 2010. Paper SPE 138241-MS.
3. Li, S.; Hendraningrat, L.; Torsæter, O. Improved oil recovery by hydrophilic silica nanoparticles suspension: 2-phase flow experimental studies. In Proceedings of the International Petroleum Technology Conference, Beijing, China, 26–28 March 2013.
4. Hendraningrat, L.; Li, S.; Torsæter, O. A coreflood investigation of nanofluid enhanced oil recovery. *J. Petrol. Sci. Eng.* **2013**, *111*, 128–138. [CrossRef]
5. Joonaki, E.; Ghanaatian, S. The application of nanofluids for enhanced oil recovery: Effects on interfacial tension and coreflooding process. *Petrol. Sci. Technol.* **2014**, *32*, 2599–2607. [CrossRef]
6. Li, S.; Torsaeter, O.; Lau, H.C.; Hadia, N.J.; Stubbs, L.P. The impact of nanoparticle adsorption on transport and wettability alteration in water-wet Berea sandstone: An experimental study. *Front. Phys.* **2019**, *7*, 74. [CrossRef]

7. Wasan, D.T.; Nikolov, A. Spreading of nanofluids on solids. *Nature* **2003**, *423*, 156–159. [CrossRef] [PubMed]
8. Li, S.; Hadia, N.J.; Lau, H.C.; Torsaeter, O.; Stubbs, L.P.; Ng, Q.H. Silica nanoparticles suspension for enhanced oil recovery: Stability behavior and flow visualization. In Proceedings of the SPE Europec featured at the 80th EAGE Conference and Exhibition, Copenhagen, Denmark, 11–14 June 2018.
9. Hiemenz, P.C.; Dekker, M. *Principles of Colloid and Surface Chemistry*, 2nd ed.; Dekker: New York, NY, USA, 1986.
10. Ghadimi, A.; Saidur, R.; Metselaar, H.S.C. A review of nanofluid stability properties and characterization in stationary conditions. *Int. J. Heat Mass Transf.* **2011**, *54*, 4051–4068. [CrossRef]
11. Chang, H.; Wu, Y.C.; Chen, X.Q.; Kao, M.J. Fabrication of Cu based nanofluid with superior dispersion. *Natl. Taipei Univ. Technol. J.* **2000**, *5*, 201–208.
12. Zhu, D.; Li, X.; Wang, N.; Wang, X.; Gao, J.; Li, H. Dispersion behavior and thermal conductivity characteristics of Al_2O_3–H_2O nanofluids. *Curr. Appl. Phys.* **2009**, *9*, 131–139. [CrossRef]
13. Wang, X.; Zhu, D.; Yang, S. Investigation of pH and SDBS on enhancement of thermal conductivity in nanofluids. *Chem. Phys. Lett.* **2009**, *470*, 107–111. [CrossRef]
14. Wei, X.; Zhu, H.; Kong, T.; Wang, L. Synthesis and thermal conductivity of Cu_2O nanofluids. *Int. J. Heat Mass Transf.* **2009**, *52*, 4371–4374. [CrossRef]
15. Sofla, S.J.D.; James, L.A.; Zhang, Y. Insight into the stability of hydrophilic silica nanoparticles in seawater for Enhanced oil recovery implications. *Fuel* **2018**, *216*, 559–571. [CrossRef]
16. Hwang, Y.; Lee, J.K.; Lee, C.H.; Jung, Y.M.; Cheong, S.I.; Lee, C.G.; Ku, B.C.; Jang, S.P. Stability and thermal conductivity characteristics of nanofluids. *Thermochim. Acta* **2007**, *455*, 70–74. [CrossRef]
17. Huang, J.; Wang, X.; Long, Q.; Wen, X.; Zhou, Y.; Li, L. Influence of pH on the stability characteristics of nanofluids. In Proceedings of the 2009 Symposium on Photonics and Optoelectronics, Wuhan, China, 14–16 August 2009; pp. 1–4.
18. Zhu, H.; Lin, Y.; Yin, Y. A novel one-step chemical method for preparation of copper nanofluids. *J. Colloid Interface Sci.* **2004**, *277*, 100–103. [CrossRef] [PubMed]
19. Selvamani, V. Stability studies on nanomaterials used in drugs. In *Micro and Nano Technologies, Characterization and Biology of Nanomaterials for Drug Delivery*; Mohapatra, S.S., Ranjan, S., Dasgupta, N., Mishra, R.M., Thomas, S., Eds.; Elsevier: Dordrecht, The Netherlands, 2019; pp. 425–444. ISBN 9780128140314.
20. Ranka, M.; Brown, P.; Hatton, T.A. Responsive stabilization of nanoparticles for extreme salinity and high-temperature reservoir applications. *ACS Appl. Mater. Interfaces* **2015**, *7*, 19651–19658. [CrossRef] [PubMed]
21. Hwang, Y.; Lee, J.; Lee, J.; Jeong, Y.; Cheong, S.; Ahn, Y.; Kim, S.H. Production and dispersion stability of nanoparticles in nanofluids. *Powder Technol.* **2008**, *186*, 145–153. [CrossRef]
22. Zhu, H.; Zhang, C.; Tang, Y.; Wang, J.; Ren, B.; Yin, Y. Preparation and thermal conductivity of suspensions of graphite nanoparticles. *Carbon* **2007**, *45*, 226–228. [CrossRef]
23. Lee, J.; Hwang, K.; Jang, S.; Lee, B.; Kim, J.H.; Choi, S.; Choi, C. Effective viscosities and thermal conductivities of aqueous nanofluids containing low volume concentrations of Al_2O_3 nanoparticles. *Int. J. Heat Mass Transf.* **2008**, *51*, 2651–2656. [CrossRef]
24. Oh, D.; Jain, A.; Eaton, J.; Goodson, K.; Lee, J. Thermal conductivity measurement and sedimentation detection of aluminum oxide nanofluids by using the 3-omega method. *Int. J. Heat Fluid Flow* **2008**, *29*, 1456–1461. [CrossRef]
25. Oh, D.; Jain, A.; Eaton, J.; Goodson, K.; Lee, J. Thermal conductivity measurement of aluminum oxide nanofluids using the 3-omega method. In Proceedings of the ASME 2006 International Mechanical Engineering Congress and Exposition, Chicago, IL, USA, 5–10 November 2006; pp. 343–349.
26. Wang, H.; Sen, M. Analysis of the 3-omega method for thermal conductivity measurement. *Int. J. Heat Mass Transf.* **2009**, *52*, 2102–2109. [CrossRef]
27. Metin, C.O.; Lake, L.W.; Nguyen, Q.P. Stability of aqueous silica nanoparticle dispersions. *J. Nanopart. Res.* **2011**, *13*, 839–850. [CrossRef]
28. Hendraningrat, L.; Li, S.; Torsæter, O. Enhancing oil recovery of low-permeability berea sandstone through optimised nanofluids concentration. In Proceedings of the SPE Enhanced Oil Recovery Conference, Kuala Lumpur, Malaysia, 2–4 July 2013.

29. Khezrnejad, A.; James, L.A.; Johansen, T.E. Nanofluid enhanced oil recovery-mobility ratio, surface chemistry, or both? In Proceedings of the International Symposium of the Society of Core Analysts, St. John's, NL, Canada, 16–21 August 2015.
30. Rusli, W.; Jackson, A.W.; Van Herk, A. A Roadmap towards successful nanocapsule synthesis via vesicle templated RAFT-based emulsion polymerization. *Polymers* **2018**, *10*, 774. [CrossRef] [PubMed]
31. Turbiscan^{LAB} manual.

 *nanomaterials*

Article

Experimental Investigation of the Effect of Adding Nanoparticles to Polymer Flooding in Water-Wet Micromodels

Edgar Rueda [1,*] , Salem Akarri [2,*], Ole Torsæter [2] and Rosangela B.Z.L. Moreno [1]

1   School of Mechanical Engineering, University of Campinas, Rua Mendeleyev, 200 Cidade Universitária Barão Geraldo, Campinas–SP CEP 13083-860, Brazil; zanoni@fem.unicamp.br
2   PoreLab Research Centre, Department of Geoscience and Petroleum, Norwegian University of Science and Technology (NTNU), S. P. Andersens veg 15a, 7031 Trondheim, Norway; ole.torsater@ntnu.no
*   Correspondence: edgar.rueda.r@gmail.com (E.R.); salem.s.f.akarri@ntnu.no (S.A.); Tel.: +55-19-981-24-8673 (E.R); +47-465-63-030 (S.A.)

Received: 20 May 2020; Accepted: 16 July 2020; Published: 29 July 2020

Abstract: Recently, the combination of conventional chemical methods for enhanced oil recovery (EOR) and nanotechnology has received lots of attention. This experimental study explores the dynamic changes in the oil configuration due to the addition of nanoparticles (NPs) to biopolymer flooding. The tests were performed in water-wet micromodels using Xanthan Gum and Scleroglucan, and silica-based NPs in a secondary mode. The microfluidic setup was integrated with a microscope to capture the micro-scale fluid configurations. The change in saturation, connectivity, and cluster size distributions of the non-wetting phase was evaluated by means of image analysis. The biopolymer content did not affect the ability of the NPs to reduce the interfacial tension. The experiments showed that the reference nanofluid (NF) flood led to the highest ultimate oil recovery, compared to the Xanthan Gum, Scleroglucan and brine flooding at the same capillary number. In the cases of adding NPs to the biopolymer solutions, NPs-assisted Xanthan flooding achieved the highest ultimate oil recovery. This behavior was also evident at a higher capillary number. The overall finding suggests a more homogenous dispersion of the NPs in the solution and a reduction in the polymer adsorption in the Xanthan Gum/NPs solution, which explains the improvement in the sweep efficiency and recovery factor.

Keywords: enhanced oil recovery; chemical flooding; biopolymer; silica nanoparticles; microfluidics

1. Introduction

One of the most common enhanced oil recovery (EOR) techniques is polymer flooding. That is described by adding polymer molecules to the aqueous injected phase aiming to increase the viscosity of the solution to be injected, reducing the mobility ratio of water to oil and to controlling the fingering effects. However, polymer waterfloods are highly probable to result in permeability reduction [1]. In addition, polymer flooding may be considered as an improved waterflooding method rather than an EOR method, since it does not ordinarily unlock the isolated residual oil by water in the porous medium [2]. The efficiency of a polymer flooding is related to three aspects: (1) the reduction in the injected fluid mobility, (2) polymer alteration of fractional flow curves, and (3) diversion of the injected water from swept areas [3].

The Xanthan Gum and Scleroglucan are two types of microbial-origin biopolymers. The first polymer is obtained from the microorganism Xanthomas Campestris [4], and the second is a class of fungal polysaccharides secreted extracellularly by fungi of the genus Sclerotium [5]. Xanthan Gum is an anionic polymer with an estimated molecular weight of 2.65 Million of Daltons, the Pyruvate and

Acetate content are 0.9%, and 3.52%, respectively [6]. The Scleroglucan is a neutral β-1, 3-β-1,6-glucan. X-ray diffraction shows a triple-helical conformation in the solid-state [7]. In aqueous solution, the Scleroglucan molecule exists in a stiff, triple-stranded helical structure, where side chains are exposed toward the exterior [8]. The Scleroglucan molecular weight is approximately 5.2 Million of Daltons [9].

The biopolymers have been applied in the North Alma Penn unit [10]. In the early stages of the waterflooding and using concentrations from 250 ppm to 304 ppm, the results showed no damage in the formation, and the decline in the water production in some production wells could be attributed to the positive effect of the biopolymers. In the case of the field Eddesse-Nord sandstone in Germany [11], the implementation of a Xanthan Gum pilot project at 800 ppm of concentration showed successful injection results, in terms of lower values of adsorption than expected and low degradation effects.

Nanotechnology has received much attention in different engineering disciplines in the oil industry. An important example is the positive impact on the oil recovery factor because of the improvement in the sweep efficiency and reduction in the trapped oil by the capillary pressure [12–14]. The four mechanisms investigated in literature as oil displacement mechanisms by NPs are disjoining pressure gradient at oil–NPs interface, the density difference between water and NPs, wettability alteration, and reduction in interfacial tension [15]. The disjoining pressure is defined as the pressure required to oppose the fluid/solid attractive forces and lift the film from a solid surface. In other words, this pressure represents the net pressure difference between the pressure in a thin film and that in the bulk liquid from which the thin film extends [16]. The density difference between the nanoparticle and the water drives the particles to agglomerate in the smallest pores and throats. That difference generates an increase in the pressure, which mobilizes the oil in the adjacent pores [17]. NPs have the ability to alter the wettability from oil-wet to neutral or more water-wet surface [15]. The use of hydrophilic silica NPs has shown a reduction in the interfacial tension between the brine and oil when the concentration of NPs increases [18,19].

In the last decade, the combination of the polymer flooding technique and NPs has been investigated as a promising method to reduce trapping efficiency. NPs (silica NPs, surface-modified silica NPs, or nanoclay) have been added to polymer solutions (hydrolyzed polaycarylamide (HPAM) or Xanthan Gum (XG)) to improve the performance of polymer waterflooding. A micromodel-study presented that silica NPs and HPAM flooding increased the oil recovery by 10% of original oil in place (OOIP) than polymer flooding due to the wettability alteration mechanism [20]. Another micro-scale study [21], presented that the efficiency of the pore-to-pore displacement of heavy oil was improved by an increase in the concentration of silica NPs in polaycarylamide (PAM) solutions, even when salt was present in the system. That was explained by ion-dipole interactions between the NPs and cations reducing the polymer degradation. Their study shows that alteration is more effective at lower polymer concentrations. However, for field applications, the use of HPAM is more common than PAM due to dilution problems of PAM. Cheraghian and Khalilinezhad, 2015 [22] tested the addition of nanoclay (at three concentrations: 0.8, 0.9, and 1.0 wt.%) to polymer flooding (at different concentrations of HPAM) on core-scale. They found that the best recipe resulted in incremental heavy oil recovery by a factor of 5% of OOIP in comparison to polymer flooding after one pore volume fluid injection. Another experimental work showed that silica NPs and nanoclay increased the PAM solution viscosity and decreased its adsorption onto the rock [23]. The core-scale work by Saha et al., 2018 [24] showed that nanoparticle assisted polymer (Xanthan Gum) flooding was effective for EOR applications in heavy crude oil systems. Based on the stability analysis of silica NPs, the NPs were more stable in the polymer solution than in water. Besides, the viscosity of the polymer solution improved, and the interfacial tension reduced after the addition of NPs. A considerable improvement in sweep efficiency was observed by adding surface-modified silica NPs to XG solutions, but not to HPAM solutions [25]. Mohammadi et al., 2019 [26] studied the effect in the rheology of polymeric solutions with silica NPs. Their results showed that the viscosity of PAM solutions increased when the concentration of NPs was respectively 0.5, 1.0, 1.5, and 2.0 wt.%. A faster increase in thickness occurred when the concentration was higher than 1.5 wt.% at 25 °C. This point corresponds to the critical nanoparticle concentration (CNC). Kennedy et al., 2015 [27] found in a rheological study that Xanthan Gum exhibited shear-thinning over the entire shear rate range and enhanced the viscosity and the storage moduli with the increasing particle concentration. This phenomenon created larger domains of the associated polymer gel as a result of the interaction between the polymer and the particles.

Microfluidics has become a vital research area in the petroleum industry. It is highly appreciated owing to significantly improving our current understanding of pore-scale displacement events and interactions occurring within tiny fluid volumes, moving within well-defined pore-structures. In addition, it enables capturing the dynamic changes within the medium with a high spatial and temporal resolution. The energy and chemicals fed to a microfluidic setup are much less compared to the core-flooding apparatus, which lowers costs and risks. The heavy and conventional core flooding setup is reduced to a smaller apparatus with a higher level of accuracy and significant observations, more tests can be performed to validated and compare different flooding scenarios and the result takes part in the selection criteria previous to more complex tests like core-flooding.

The ratio of the viscous forces to the capillary forces defined as the capillary number (Nc), see Equation (1), indicates what forces dominate the flow regime, viscous or capillary flow. Where V_w is the interstitial velocity, μ is the viscosity and σ_{ow} is the interfacial tension between the fluids involved in the displacement. In conventional brine flooding, the increase in the injection rate (higher capillary number) can reduce the trapped oil in the reservoirs and glass-bead packs [28,29]. In the case of fluids with NPs, the tendency is different because the NPs require time to modify the wettability of the surface by disjoining pressure [29] or because at higher rates the particles agglomerate in the porous media [13,29]:

$$N_c = \frac{V_w\,\mu_w}{\sigma_{ow}}. \tag{1}$$

The water injectivity (I_w) during a water flooding process through a porous media is proposed by Civan, 2016 [30] in Equation (2) where q_w is the water rate and ΔP is the pressure difference across the porous medium:

$$I_w = \frac{q_w}{\Delta P}. \tag{2}$$

The injectivity index was used in this work to evaluate the effect of each solution during the displacement, according to a reference solution (brine in the case of the fluids without NPs and NPs in the case of the solution with silica NPs). That is calculated using Equation (3):

$$I_{index} = \frac{I_{polymer}}{I_{reference}}. \tag{3}$$

Modeling these phenomena is challenging. Simulation models need to take into consideration a multi-phase multi-component interactive system. It is necessary to represent the effects of the nanoparticles on the interfacial tension, capillary pressure, and wettability. Polymer flooding also includes interactive phenomena. The non-Newtonian behavior of the polymer solutions requires local and instantaneous determination of the displacing fluid viscosity as a function of shear rate and polymer concentration. However, retention and degradation phenomena strip polymer from the solution, which justifies the multicomponent approach. Changes in local permeability can take place because of polymer adsorption. All those phenomena have recently received close attention, and semi-analytical procedures [31] or commercial simulators [32] are progressively coupling their effects on chemical enhanced oil recovery. Nevertheless, the associative contributions of nanoparticles and polymer is new, opening opportunities for developments in this area.

In this work, we study the improvement in the sweep and the displacement efficiencies by the polymer flooding using Xanthan Gum and Scleroglucan assisted with silica NPs. Both phenomena were evaluated independently and simultaneously from two-dimensional (2D) image analysis obtained during microfluidics tests in water-wet glass micromodels and translated in terms of oil recovery factor and injectivity loss. The combination of both chemicals represents an environmentally friendly alternative to improve conventional water flooding. This study also provides new insight into the application of NPs in the oil and gas industry

2. Materials and Methods

2.1. Biopolymers, NPs, and Brine

Two different types of dry based biopolymers; Xanthan Gum and Scleroglucan were diluted in deionized water to reach a concentration of 0.4 wt.%. To have full hydration of the polymers, they were mixed using a magnetic stirrer for 24 h for the Xanthan Gum and seven days for the Scleroglucan at room temperature. Those solutions are the stock to prepare the dilutions.

The silica nanoparticle used in this study is a special laboratory research development for Evonic. The name of AEROSIL® originally markets the product. The main component was silicon dioxide with other minor parts of alumina. The provided particles have attached chains of polymers to their surfaces to ensure long-term stability in solution. This type of nanomaterial is known as polymer-coated silica nanoparticles. The solution in gel was supplied as by the name AERODISP®. The charge of this type of nanoparticle is anionic equal to the Xanthan Gum, while Scleroglucan molecule is neutral. The properties of the NPs are given in Table 1. The 3 wt.% sodium chloride brine was prepared using a magnetic stirrer for 30 min in deionized water. The density was 1.019 g/cm^3, the viscosity of the solution was 1.04 mPa·s and the pH is 6.52. All the values were measured at room temperature (22 °C).

Table 1. Properties of the stock liquid surface-modified silica NPs solution.

Properties	Value
NPs concentration	26 wt.%
Basis	SiO$_2$ (sol-gel-anionic)
Modification	Polymer
Average Size	32 nm
Solvent	Deionized water
Specific surface area	140–220 m^2/g

Three groups of dilutions were prepared from the stock solutions. The first group is composed of the brine and the 0.1 wt.% NPs solution. In the second group, two polymeric solutions were prepared using biopolymers, Xanthan Gum or Scleroglucan, and in the last group, at each polymeric solution, NPs were added. Overall, six solutions were prepared in the concentrations shown in Table 2. Each solution was mixed for 30 min using a magnetic stirrer at 500 rpm and the solutions were stored at 4 °C.

Table 2. Solutions composition.

Solution Type	Sample Name	Polymer Concentration (wt.%)	NPs Concentration (wt.%)	Brine Concentration (wt.%)
Brine	Brine	0	0	3.00
Nanoparticle	NF	0	0.100	3.00
Xanthan Gum	XG	0.016	0	3.00
Scleroglucan	SCL	0.025	0	3.00
Xanthan Gum/NP	XG_NP	0.016	0.100	3.00
Scleroglucan/NP	SCL_NP	0.025	0.100	3.00

2.2. Crude Oil

Dead oil from the North Sea was used in this study. The filtration process was performed using a filter whose pore size was 1.1 µm. The density and viscosity measured at 22 °C were 0.892 g/cm^3

(27 °API), and 21.9 mPa·s, respectively. Saturate, aromatic, resin and asphaltenes (SARA) analysis showed that the oil contains 71.57% of saturates, 20.81% of aromatics, 7.44% of resins and 0.18% of asphaltenes.

2.3. Glass Micromodel

Borosilicate glass micromodels chips were used in the study (Figure 1). Those models represent a real porous media network, and under a microscope, due to the clear optical properties from all sides, allowing the understanding of the flow dynamic behavior of the wetting and non-wetting phases in a displacement process. The size of the chip was 45 mm × 15 mm and the thickness was 1800 μm. The dimensions of the pore-network were 20 mm × 10 mm and the thickness was 20 μm. According to the manufacturer specifications for the chip, the pore volume was 2.3 μL, the porosity was 57% and the permeability was 2.5 Darcy. The porosity value was also confirmed by the image analysis. Single-phase flooding tests were performed to estimate the liquid permeability of the pore-network according to the work presented by Pradhan et al., 2019 [33]. The network permeability was around 8.3 Darcy.

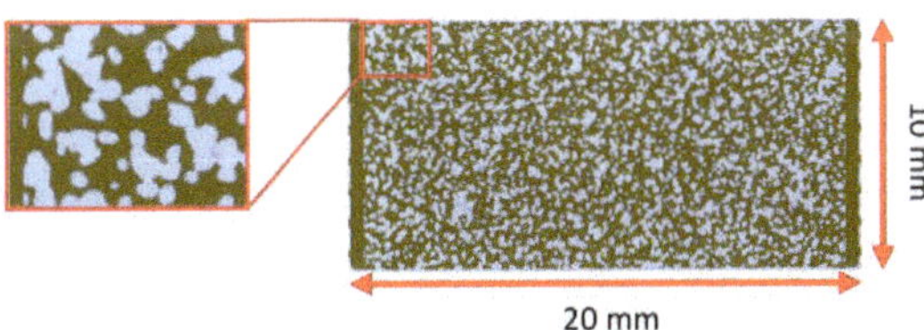

Figure 1. Pore-network of the glass micromodel used in this study (Blue: grain area; brown: pore area).

2.4. Micromodel Setup

The micromodel was attached to an aluminum platform of 128 mm × 85.4 mm × 20 mm that operates at a maximum of 10 bar of pressure and 80 °C. The inlet and outlet ports are connected using Teflon tubing. The visualization was done under a microscope equipped with a digital camera used to capture images every 30 s. The fluids were injected using a syringe pump. A pressure transducer was adapted in the inlet and outlet lines to record the pressure during the recovery agent injection. Figure 2 shows a representation of the micromodel set up.

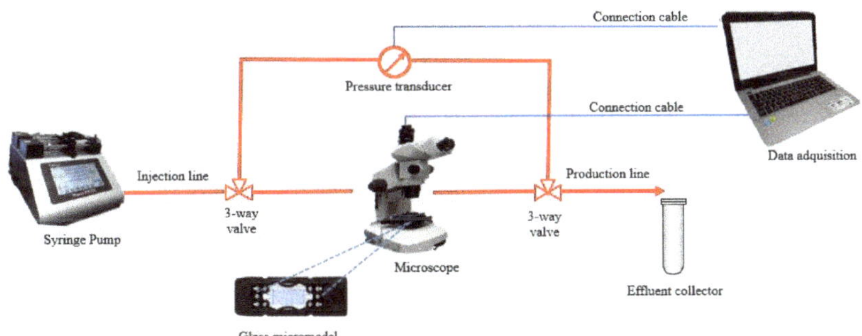

Figure 2. Diagram of the micromodel flooding setup.

2.5. Interfacial Tension and Contact Angle

The interfacial tension test was made using a drop shape analyzer and using a J-shaped needle with a diameter of 1.001 mm. The fluid container dimensions are 30.0 mm length, 20.0 mm width and 30.6 mm height, at room conditions. The pendant drop method was used to analyze the drop shape of oil against the aqueous phase and with the Young-Laplace equation (Equation (4)) to calculate the interfacial tension.

$$\sigma = \frac{\Delta\rho \cdot g \cdot R_o}{\beta} \tag{4}$$

where:

σ = interfacial tension.

g = gravitational constant.

R_o = radius of drop curvature.

β = shape factor.

$\Delta\rho$ = density difference

The measurement was performed for all solutions for 2 h to reach the equilibrium condition and it was repeated to confirm the result.

The contact angle between a drop of crude oil and the polymer solutions was measured using the same apparatus. The solid surface used was glass and the oil drop was placed under the surface (captive bubble method).

2.6. Nanosize Distribution

The particle size distribution is measured using a Dynamic Light Scattering (DLS) instrument. The apparatus is a Zetasizer Nano series, where the diameter of the sphere that diffuses in the solution is measured from the Brownian motion of the particles in a sample. DLS and established theories were used to determine the particular size distribution and the relationship with the diffusion speed.

2.7. Microfluidics Tests

All tests were developed using a secondary methodology to evaluate the effect of different recovery agents in terms of oil recovery factor, pressure drop and sweep efficiency. The chips were cleaned using toluene, methanol, acetone, and distilled water and then dried in an oven at 60 °C overnight. The brine saturation was performed under vacuum (~80 mTorr). The brine was injected at different rates to push out any remaining air in the chip.

The oil injection was performed at 1, 10, 20, 30, 70 and 100 µL/min until no more water was produced. Under this condition, the remaining brine in the chip is the connate water and the oil saturation is the maximum. The microfluidic tests were divided into three parts. The first four tests were performed at a capillary number of 1×10^{-6} in order to evaluate the biopolymer solutions in comparison with the brine at the same flow regime. In the second part, two more experiments were developed at the 2×10^{-6} capillary number. In this case, Xanthan Gum solutions assisted with the NPs and the reference nanofluid (NF) were studied. Finally, the next two tests evaluated the reference NF and the Xanthan Gum solution with NPs at a higher capillary number by increasing the flow rate to 0.39 µL/min to see the effect of a high rate injection on the performance of the solutions. The recovery agent was injected until no more oil was produced. Due to the differences in the performance of each solution, different amounts of fluid are necessary to reach steady-state conditions in each test. The dynamic study was made taking images every 30 s during the recovery agent injection and the automatic image analysis was done using a Matlab routine. The time was reported as porous volumes injected (PVI). This was calculated from the porous volume of the entire chip. However, the image analysis was performed in a representative area located in the center of the chip, which is 40% of the total area. The segmentation uses the color thresholding method to extract the oleic phase. That results in binary images were analyzed to evaluate the total area of oil and the number, size, and distribution of the clusters as the function of the time and the characteristic Euler number. According to [34], for a 2D image, the Euler number is defined as the number of connected components—the number of holes. Figure 3a,b show the initial condition Swi (initial water saturation) and the final condition Sor (residual oil saturation) from the original images of the micromodel, where the oleic phase corresponds to the yellow part. The images after the treatment are shown in Figure 3c,d, where the white area corresponds to the oleic phase and in black refers to the porous media, brine and recovery agent.

Original   **Segmented**

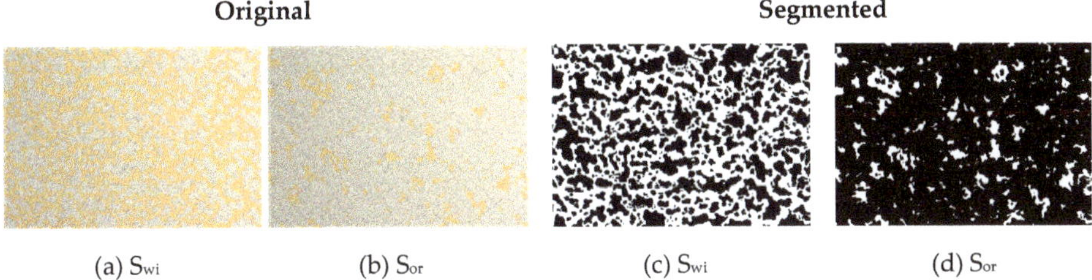

(a) S_{wi} (b) S_{or} (c) S_{wi} (d) S_{or}

Figure 3. Original and segmented images of the pre- and post-flooding state in the microchip: (**a**) initial water saturation—original; (**b**) residual oil saturation—original; (**c**) initial water saturation—segmented; (**d**) residual oil saturation—segmented.

3. Results

3.1. Fluids Characterization

The use of polymeric solutions has the objective of increasing the viscosity of the solution to be injected to reduce the mobility ratio between the oil and the injected fluid. In this study, the biopolymers increased the viscosity of the brine more than two times. A little additional increase in thickness was reached when the NPs were added, except for the Xanthan Gum where the viscosity reduced by 2.65%. The NPs increased the pH of the solution on an average of 6.21%. Density measurements were based on the oscillation U-tube method. No considerable changes in the results of density measurements of the aqueous solutions were observed after adding NPs. The measured values of the viscosity and the density of the filtered crude oil sample from the North Sea (Dead Oil) were 21.9 mPa·s, and 0.892 g/cm^3, respectively. The results are shown in Table 3.

Table 3. Density, viscosity, pH and nano-size distribution results of the fluids used in this study.

Fluid	Density (g/cm^3)	Viscosity (mPa·s)	pH	Nano-size Distribution
Brine	1.019	1.04	6.52	-
NF	1.021	1.07	6.72	64.82
XG	1.020	2.26	6.00	-
SCL	1.020	2.44	6.84	-
XG_NP	1.021	2.20	6.60	31.87
SCL_NP	1.021	2.46	7.22	31.48
Dead_Oil	0.892	21.9	-	-

3.2. Nano-Size Distribution

According to the supplier of the silica-NPs, the average size of the NPs is 32 nm. Even so, measurement of nano-size distribution was conducted to confirm the distribution of the NPs into the solutions (Table 3). The results showed 64.82 nm for the NF solution without biopolymers. This value is higher than that reported by the supplier because some particles could be agglomerated. That difference on the agglomeration is not significate, but besides that, the measurements were performed under static conditions. Therefore, agglomeration with higher nano-size distribution could happen under dynamic flow conditions through the porous media. This behavior is probably controlled when the biopolymers is added in the solution, which showed values for the average size of 31.87 nm for XG_NP and 31.48 nm for SCL_NP. This result could indicate a better distribution of the NPs when the biopolymers are in the solutions.

3.3. Interfacial Tension and Contact Angle

The addition of hydrophilic silica NPs into the brine-oil system reduced the interfacial tension (IFT). Comparing the brine (Figure 4a) with the NF solution (Figure 4b), the IFT of the oil-NPs solution is one-third of the oil-brine system. When the biopolymer is present in the solution, the interaction between the polymers and the NPs did not affect the reduction due to NPs. The polymers slightly improved the reduction effect in the interfacial tension.

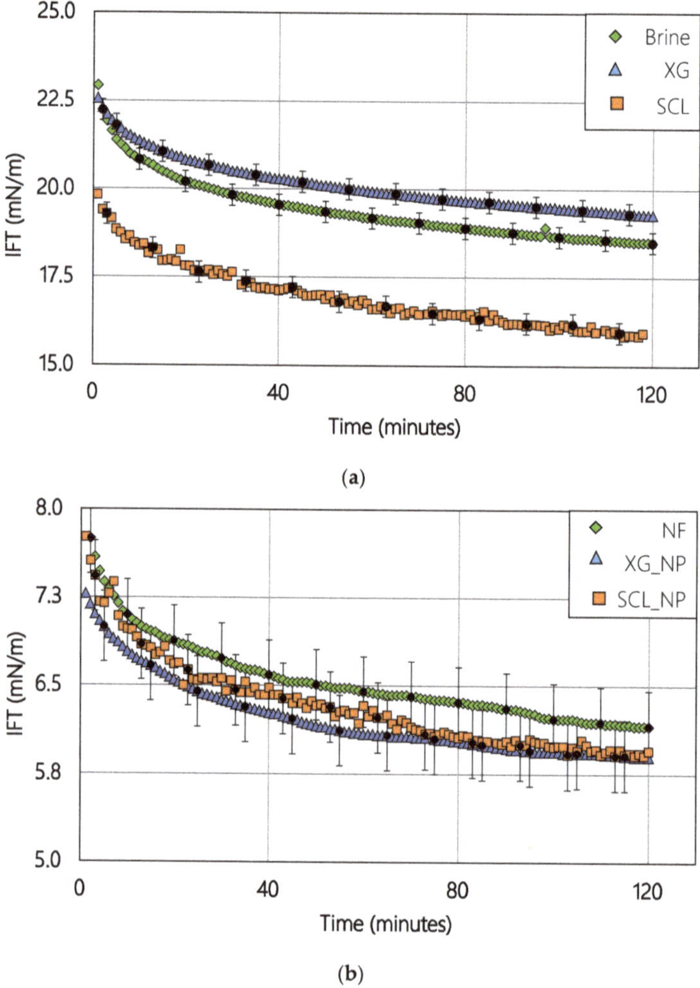

Figure 4. (a) Interfacial tension-free of NPs and (b) Interfacial tension in solutions with NPs.

In the case of the contact angle, the behavior is similar to the IFT, where the NPs at 0.1 wt.% reduced the contact angle by 36% in comparison with the brine. However, when the biopolymers are in the solution, the reduction in the contact angle was less evident, showing values within 25% on average. Those results are shown in Figure 5.

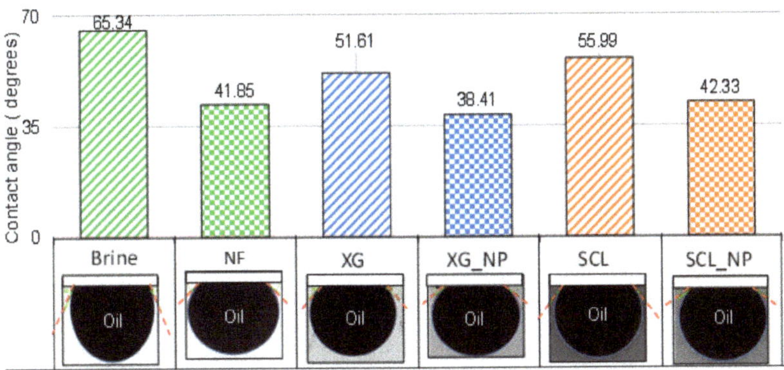

Figure 5. Results of the contact angle measurement for the fluids used in this study.

3.4. Microfluidic Experiments

The screening process was made as a function of the type of fluids. For this reason, the flow regime was maintained constant based on the concept of capillary number. The porous media and the oil phase were the same for all tests. Secondary injection methodology and water-wet glass micromodels were used in the microfluidics tests. The initial water saturation was established by using a brine with 3 wt.% sodium chloride in all experiments. The same methodology was performed for brine and oil saturation to have the same initial condition previous to the recovery agent injection. The validation of the methodology was done performing arbitrarily the third test (SCL) more than one time. The results showed relative difference of 2.87% in the ultimate oil recovery factor at 3.4 PVI. That difference was considered as acceptable for the objective of the paper and was used as an argument to validate the microfluidic laboratory methodology used in this work.

3.4.1. Microfluidic Screening of Biopolymer Solutions

In the first part of the study, three microfluidics tests were performed. The capillary number was constant at 1×10^{-6} to have the same flow regime in the porous medium. The flow rate for each test is shown in Table 4. In this case, brine with the biopolymers Xanthan Gum and Scleroglucan were evaluated.

Table 4. Biopolymer and NPs solutions flow rate selection at 1×10^{-6} capillary number.

Test No.	Recovery Agent	IFT (mN/m)	Flow Rate (µL/min)	Capillary Number	S_{wi}	Ultimate Oil Recovery (frac. of OOIP)
1	Brine	18.58	0.121	1×10^{-6}	0.13	0.76
2	XG	19.38	0.058	1×10^{-6}	0.16	0.77
3	SCL	16.00	0.044	1×10^{-6}	0.20	0.69

The biopolymers solutions as a recovery agent were studied and compared with conventional water flooding. According to Figure 6a, the Scleroglucan showed poor behavior obtaining the lowest oil recovery factor (69%) and the most delayed equilibrium point (3.8 PVI). The second biopolymer, Xanthan Gum, reached the same recovery as the brine (76%). However, the production with the biopolymer injection reaches the steady-state after the first porous volume injected, while the brine takes 2.4 PVI to reach the equilibrium. In terms of the number of oil clusters (Figure 6c) and oil connectivity (Figure 6b), the remaining number of clusters, the results for Xanthan Gum and Scleroglucan were the same, but the connectivity between those oil bodies was higher for the Scleroglucan. Finally, Figure 6d shows the relation between the number and size of the clusters. The tendency is similar for the brine and the

Xanthan gum. In the case of the Scleroglucan, the graph indicates bigger remaining oil clusters after reaching the steady-state flow condition. Table A1 (Appendix A) shows the segmented images of the chip during the injection for each solution as a function of the porous volume injected (PVI).

The pressure response during the recovery agent injection is shown in Figure 6e. The results have a concordance with the flow rate, and in consequence, higher rates correspond to higher differential pressure. The hydrodynamic resistance during polymer flooding generate a pressure increase during the injection of the fluids without NPs. In terms of the injectivity index, Figure 6f shows a reduction in the injectivity when the biopolymers are injected in comparison with the brine injection.

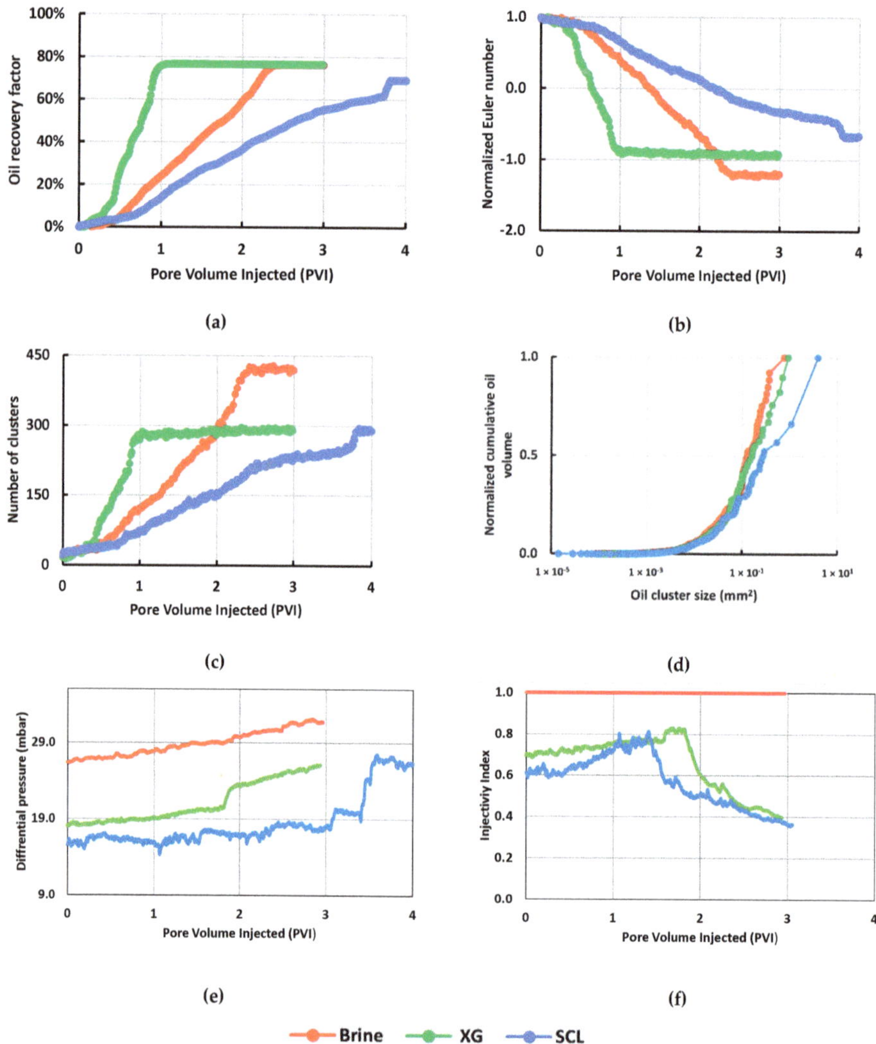

Figure 6. Microfluidic results for screening of biopolymer solutions at $N_c = 1 \times 10^{-6}$: (**a**) Oil recovery factor; (**b**) Normalized Euler number; (**c**) Number of clusters; (**d**) Normalized cumulative oil volume; (**e**) Differential pressure; (**f**) Injectivity index.

3.4.2. Effect of Adding NPs to Xanthan Gum Solutions

In the second part of the study, two more microfluidics tests were performed to evaluate the effect of the NPs and biopolymer as recovery agents. The capillary number was 2×10^{-6}. The details for each test are shown in Table 5. NPs-assisted Scleroglucan flooding was not successful for the investigated application due to significant injectivity problems.

Table 5. Biopolymer and NF flow rate selection at 2×10^{-6} capillary number.

Test No.	Recovery Agent	IFT (mN/m)	Flow Rate (µL/min)	Capillary Number	S_{wi}	Ultimate Oil Recovery
4	NF	6.19	0.078	2×10^{-6}	0.17	0.84
5	XG_NP	5.92	0.039	2×10^{-6}	0.14	0.88

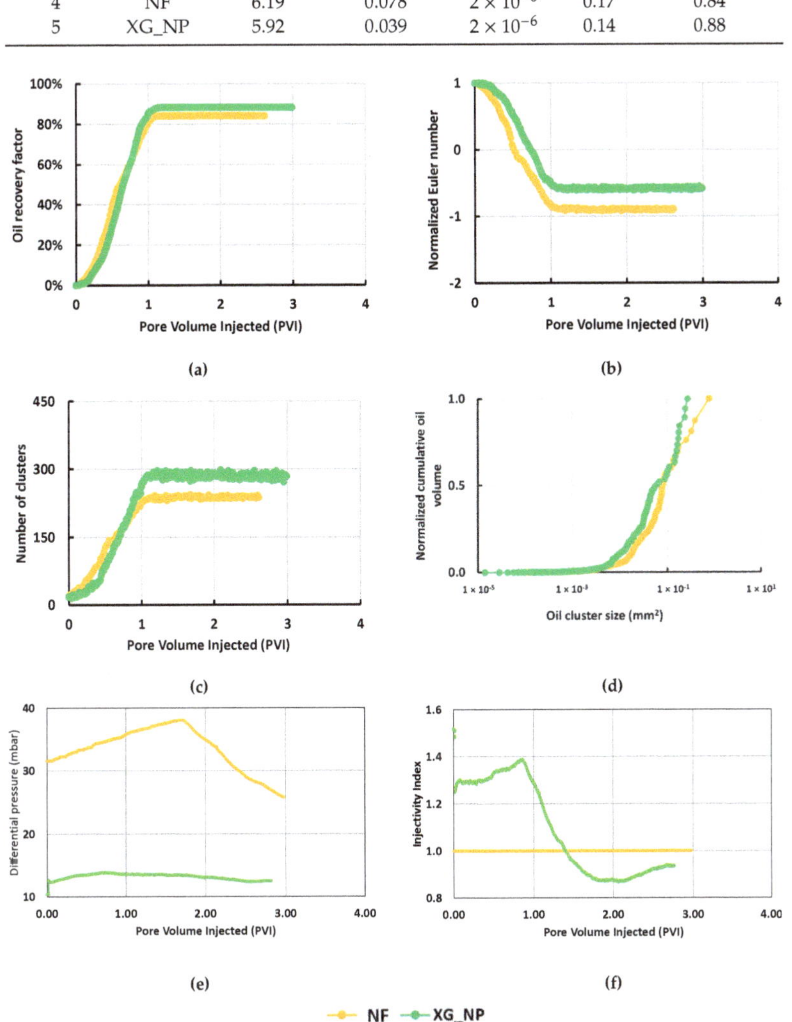

Figure 7. Microfluidic results showing the effect of adding NPs to xanthan gum solutions at $N_c = 2 \times 10^{-6}$: (**a**) Oil recovery factor; (**b**) Normalized Euler number; (**c**) Number of clusters; (**d**) Normalized cumulative oil volume; (**e**) Differential pressure; (**f**) Injectivity index.

The effect of the NPs in the Xanthan Gum solution was studied as a function of the oil recovery factor and the non-wetting clusters. Figure 7a showed an increase in the ultimate recovery from 84% for the NF solution to 88% for the Xanthan Gum solution with the silica NPs. For both fluids, the steady-state is reached after the injection of one pore volume. According to Figure 7b,c, the number of oil clusters and the connectivity was lower to the NF system compared to the biopolymer/NP system. However, Figure 7d showed that the size of the clusters is smaller for the solution with Xanthan Gum and for this reason, the recovery is higher. The segmented images used in the calculus are shown in Table A2 (Appendix A) as a function of the time (PVI) for each recovery agent.

In the case of differential pressure, hydrodynamic resistance generates a pressure increase. However, after the end of the oil production, the pressure decreases due to the difference in the fluid viscosity. Figure 7e shows that the ΔP for XG_NP decreased to a third compared to the NF fluid. The relation between the injection rate and the differential pressure in terms of the injectivity index (Figure 7f) shows an improvement in the injectivity of the system biopolymer-NPs in comparison with the solution without Xanthan Gum at $N_c = 2 \times 10^{-6}$.

3.4.3. Effect of the Flow Rate on the Recovery Performance

In the third part of the study, two more microfluidic tests were performed at a high and constant rate to evaluate the effect of the biopolymer on the NPs behavior in an unfavorable flow regime scenery for the NPs. The properties of each test were shown in Table 6.

Table 6. Biopolymer and NF flow rate selection of 0.39 μL/min.

Test No.	Recovery Agent	IFT (mN/m)	Flow Rate (μL/min)	Capillary Number	Swi	Ultimate Oil Recovery
6	NF	6.19	0.39	1×10^{-5}	0.18	0.78
7	XG_NP	5.92	0.39	2×10^{-5}	0.14	0.85

In Figure 8a, the Xanthan Gum with the silica reached a higher recovery (85%) in comparison with the solution free of the biopolymer (78%). Additionally, the steady-state was reached faster for the biopolymer/NPs system. The study of the clusters in Figure 8b,c shows a higher number of clusters and higher connectivity for the XG_NP recovery agent. Nevertheless, according to Figure 8d, the size of the oil bodies was bigger in the NF flooding. In consequence, the biopolymer improved the sweep efficiency. Those graphs were built using the segmented images shown in Table A3 (Appendix A).

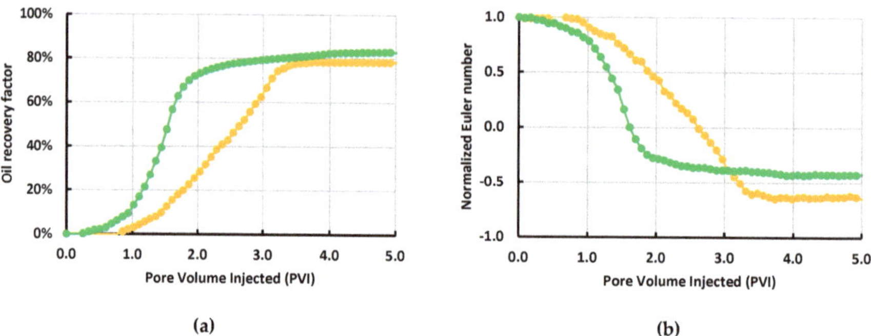

(a)

(b)

Figure 8. *Cont.*

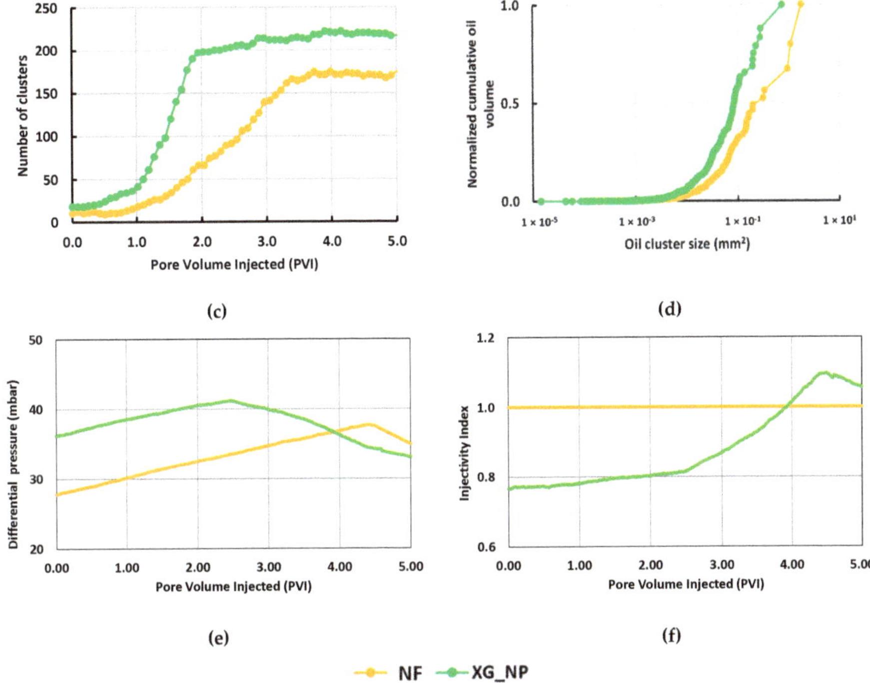

Figure 8. Microfluidic results showing the effect of a higher flow rate (0.39 µL/min) on the recovery performance: (**a**) Oil recovery factor; (**b**) Normalized Euler number; (**c**) Number of clusters; (**d**) Normalized cumulative oil volume; (**e**) Differential pressure; (**f**) Injectivity index.

Finally, the pressure response is shown in Figure 8e. Higher values of differential pressure for the XG_NP due to the viscosity increase of the biopolymer solution are observed. According to Figure 8f, Xanthan Gum, at a high rate, loses injectivity relative to the NF solution during the first 4 PVI. Thereafter, the injectivity for the system with Xanthan Gum is better than the other fluid. The effect of the hydrodynamic resistance is similar to the previous test. However, the inflection point is delayed because of the end of the oil production was delayed too.

4. Discussion

4.1. Performance of the Xanthan Gum and Scleroglucan

The performance of the solutions used in this study was evaluated comparing the reference fluids (Brine and NF) and the biopolymer solutions with NPs. In this way, were tested the main properties of the polymers and NPs. The main objective of adding polymers to the brine was to increase the viscosity to improve the sweep efficiency of the water flooding. Table 7 showed an increase of more than two times the viscosity of the biopolymer solutions in comparison with the brine. Additionally, the interaction with the Silica NPs increased this measure in 0.1 mPa·s in the Scleroglucan and the reference fluid (NF). In the xanthan gum solutions, the NPs reduced the viscosity in 0.1 mPa·s. However, no evidence of relevant polymer degradation was shown when the NPs were added to the polymer solution.

The interfacial tension reduction was a property of the NPs solutions [18,19]. The results of Table 7 suggested a reduction of 13.5 mN/m for the solutions with Xanthan Gum, in consequence, the reduction was improved in 1.1 mN/m in comparison with the reference fluids (NF and brine). On the other hand, the reduction was 10.1 mN/m for the Scleroglucan. That reduction was lower than the value reached with the reference fluids (12.4 mN/m). The interaction solid-liquid was evaluated measuring the contact angle. The behavior of the biopolymer in the reduction was less evident than the reference fluids.

Table 7. Effect of the NPs in the performance of the biopolymer solutions at room temperature.

Variable	Reference			Xanthan Gum			Scleroglucan		
	NF	Brine	Δ	XG_NP	XG	Δ	SCL_NP	SCL	Δ
Interfacial tension (mN/m)	6.2	18.6	12.4	5.9	19.4	13.5	5.9	16.0	10.1
Contact angle (degrees)	41.9	65.3	23.4	38.4	51.6	13.2	42.3	56.0	13.7
Viscosity (mPa/s)	1.1	1.0	0.1	2.2	2.3	0.1	2.5	2.4	0.1
pH	6.7	6.5	0.2	6.6	6.0	0.6	7.2	6.8	0.4
Nanosize distribution (nm)	64.8	-	-	31.9	-	-	31.5	-	-

The stability of NPs solutions is a key point because the tendency of agglomeration precipitation. The measured of 64.8 nm (approximately two times the size reported by the manufacture) in the nanosize distribution for the reference NF solution is an indication of agglomeration. However, the results of this study suggested an improvement in the NPs stability and distribution due to the biopolymers presence in the solutions. That because the values obtained (Table 7) are very close to the real size of the particles (32 nm). The interaction between the NPs and the biopolymer generates a more alkaline solution according to the increase in the pH (Table 7).

4.2. Evaluation of the Recovery Factor

The silica NPs were added in two different types of biopolymers (xanthan gum and Scleroglucan). The effect of the recovery factor was evaluated using microfluidics tests. The capillary number was maintained low (1×10^{-6} and 2×10^{-6}) to allowed the positive effect of the NPs in the reduction of the interfacial tension and the contact angle [15,18,19]. The biopolymer Scleroglucan showed poor behavior reaching the lowest (69%) and more delayed (3.8 PVI) recovery. When the NPs were added to the SCL solution, some injectivity problems were present. As a consequence, the positive results showed in the IFT and contact angle were not evaluate in microfluidic tests. In the case of the brine, the NPs improve the performance of the water flooding, increasing the recovery factor from the OOIP in 7.5% and anticipating the production reaching the steady-state 1.3 PVI early (Figure 9). The biopolymer xanthan gum showed the best performance when the NPs was added. The improvement in the recovery factor was 11.7% from the OOIP (Figure 9). In terms of time, the NPs did not change the time for reaching the steady-state. In the studies development by Maghzi et al., 2011; Maghzi et al., 2014; Cheraghian and Khalilinezhad, 2015 [20–22], similar behaviors were observed in the evaluation of polymer flooding with NPs.

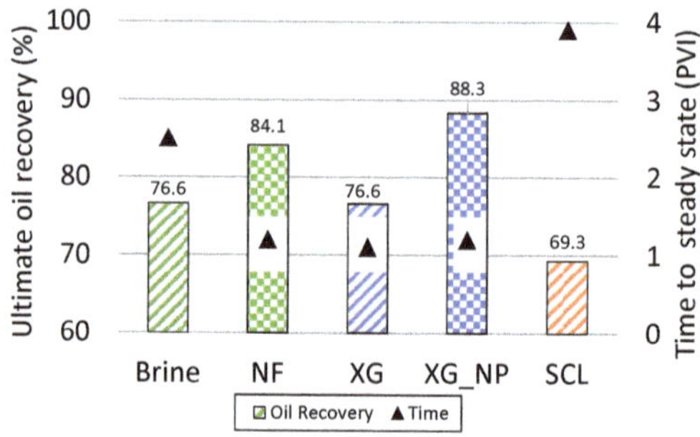

Figure 9. Ultimate oil recovery factor for the solutions evaluated using microfluidic tests as a function of the time to reach the steady-state.

4.3. NPs Effect in the Injectivity

The injectivity index was used to study the relationship between differential pressure and flow rate. The occurrence of loss of injectivity could indicate a reduction in the permeability of the porous media. In the first part of this study, the Xanthan Gum and Scleroglucan followed a similar behavior during the flooding process in comparison with the brine flooding. Still, the injectivity for both biopolymers solutions was lower than that for brine. The biopolymer adsorption could be the cause of this reduction.

In the case of the solution of Xanthan Gum with NPs, the injectivity was improved during the first 1.5 PVI. That could be due to the nanosize distribution, since a more homogeneous distribution of the NPs takes place when the biopolymer is in the solution. As a consequence, the combination of factors like interaction polymer–NPs and a low rate are controlling the polymer adsorption and NPs agglomeration, resulting in a decreased differential pressure and improved injectivity.

In a high rate scenario, the result suggested Xanthan Gum's physical adsorption on the surface of the porous media. As a consequence, permeability is reduced and the differential pressure increases. However, after some time, the agglomeration of the NPs in the fluids without polymers affects the injectivity index and showed higher values for the biopolymer system at the end of the flooding process.

4.4. Effect of the Flow Rate in the Recovery Agent Performance

The results for the ultimate oil recovery in the reference nanofluid (NF) and the xanthan gum with NPs (XG_NP) in two flow rate scenarios were shown in Figure 10. The time (PVI) required to reach this recovery was also shown. For both solutions, the ultimate oil recovery factor was lower at a high rate in comparison with the results at a low rate (2×10^{-6} capillary number). Nevertheless, the high flow rate (0.39 µL/min) affected in the same proportion the performance of both solutions. The reduction in the ultimate oil recovery was 5.7% from the OOIP for the NF and XG_NP. In terms of time, at a high rate, the steady-state point was approximately three times more delayed. As a result, the improved performance of the NPs, when the xanthan gum was in the solution, was evident even in an adverse flow regime for the NPs. The results align with the observation made by Hendraningrat et al., 2013; Zhang et al., 2016 [13,29] which explains the negative impact of high flow rates in the performance of the NPs on the oil displacement.

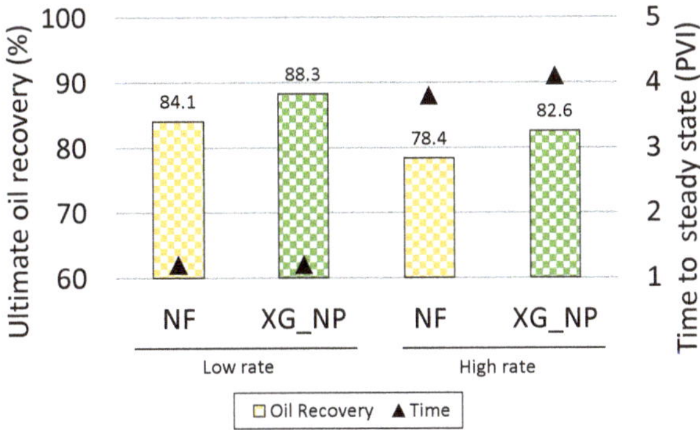

Figure 10. Ultimate oil recovery factor for the xanthan gum and NF at a low and high injections rate as a function of the time to reach the steady-state.

5. Conclusions

This study was proposed to investigate the effect of adding silica NPs in two commercial biopolymers solutions. IFT, contact angle and nanosize distribution were measured to show the effect on the fluid-fluid and fluid-solid interactions. Moreover, the solutions were tested in water-wet micromodels to investigate the effect on the sweep efficiency and displacement efficiencies in terms of oil recovery factor, and injectivity, in addition to the effect of flow rate on the recovery. Based on the results obtained, the following conclusions can be drawn:

- The interaction between the NPs and the polymer had a minor effect on the viscosity of the polymeric solution and of the brine. When the NPs were added to the Scleroglucan, a slight increase in viscosity was reached. In the case of Xanthan Gum, the interaction with the silica NPs produced a very small decrease in viscosity.

- The Xanthan Gum and Scleroglucan did not affect the reduction in the interfacial tension and contact angle of NPs solutions. Additionally, the biopolymers prevented the agglomeration of the NPs in the solution.

- The Xanthan Gum flood led to faster and higher ultimate oil recovery, smaller remaining oil clusters, and less non-wetting phase connectivity in comparison with a conventional water flooding. However, the NPs-assisted Xanthan flooding achieved the highest ultimate oil recovery and the smallest oil clusters. Additionally, at a higher rate, the system Xanthan Gum/NP showed better performance than the NF.

- In the scenario where the Xanthan Gum and the NF systems were tested at 2×10^{-6} capillary number, the injectivity improved, due to a greater homogenous dispersion of the NP in the solution and the reduction in the polymer adsorption. In the case of a higher rate, the flow of the system Xanthan Gum/NP improved the injectivity after 4 PVI.

Overall, the Xanthan gum performance improved when the NPs were in the solution in comparison with the reference fluid. These results suggested an improvement in the sweep and displacement efficiencies at the same time. Opposite, the behavior of the Scleroglucan was poor in comparison with the brine and unsuccessful in the mix with NPs. In general, this work shows new research alternatives to improve oil recovery using fluid characterization and micromodel approaches.

Author Contributions: Conceptualization, E.R., and S.A.; methodology, E.R., and S.A.; formal analysis, E.R., S.A., R.B.Z.L.M., and O.T.; investigation E.R., and S.A.; data and images curation, E.R., and S.A.; writing—original draft preparation, E.R., and S.A.; writing—review and editing, E.R., S.A., R.B.Z.L.M., and O.T.; supervision, R.B.Z.L.M., and O.T. All authors have read and agreed to the published version of the manuscript.

Funding: This research was carried out in association with the ongoing project registered as Brazilian-Norwegian Subsea Operations Consortium (BN-SOC), sponsored by Norwegian University of Science and Technology (NTNU), Norway and University of Campinas (UNICAMP), Brazil. This project is carried out in partnership with PoreLab Center of Excellence (grant number 262644), Department of Geoscience and Petroleum, Norwegian University of Science and Technology (NTNU), Norway.

Acknowledgments: This research was carried out in association with the ongoing R&D project registered as ANP 20359-6, "Injeção de Biopolímeros para a Recuperação Avançada de Petróleo de Reservatórios do Pré-Sal Brasileiro" (Universidade de Campinas (Unicamp)/Shell Brasil/ANP)–Biopolymer Injection for Enhanced Oil Recovery in Brazilian Pre-Salt Reservoirs, sponsored by Shell Brasil under the ANP R&D levy as "Compromisso de Investimentos com Pesquisa e Desenvolvimento".

Conflicts of Interest: The authors declare no interest conflicts.

Abbreviations

$^{\circ}API$	Oil relative density
CNC	Critical nanoparticles concentration
D	Darcy
EOR	Enhanced Oil Recovery
g	Gravitational constant
HPAM	Hydrolyzed polyacrylamide
IFT	Interfacial tension
I_W	Water injectivity
I_{index}	Injectivity index
$I_{polymer}$	Polymer injectivity
$I_{reference}$	Reference fluid injectivity

Nc	Capillary number
NF	Reference nanofluid
NPs	Nanoparticles
OOIP	Original oil in place
PAM	Polyacrylamide
PVI	Pore volume injected
q_w	Water rate
Ro	Radius of drop curvature
SARA	Saturate, aromatic, resin and asphaltenes
SCL	Scleroglucan
Sor	Residual oil saturation
Swi	Initial water saturation
V_w	Interstitial velocity
XG	Xanthan Gum
ΔP	Pressure difference
$\Delta \rho$	Density difference
σ_{ow}	Interfacial tension
μ_w	Water phase viscosity
β	Shape factor

Appendix A. Segmented Images from Microfluidic Tests

Binary images after the segmentation process from the images of the microfluidic tests. The images were used the calculate the behavior of the oleic phase (white) during the time for each recovery agent in the three different scenarios of the study.

Table A1. Images segmented using the Matlab code in the function of the porous volume injected (PVI) during the recovery agent injection. The fluids are the brine, xanthan gum, and Scleroglucan for the microfluidic screening of biopolymer solutions.

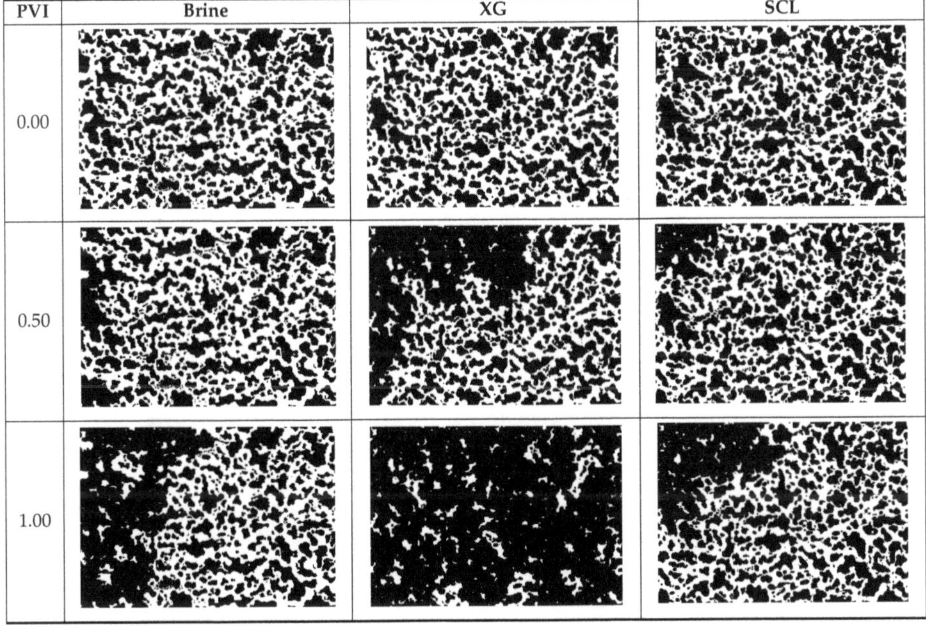

PVI	Brine	XG	SCL
0.00			
0.50			
1.00			

Table A1. *Cont.*

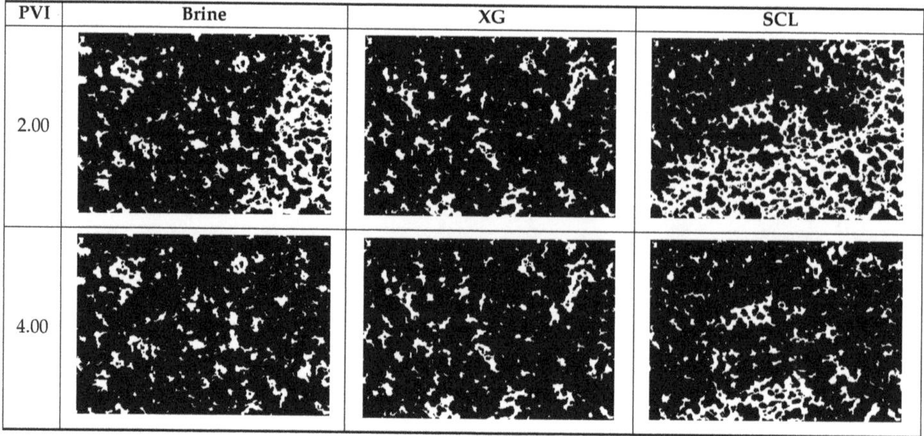

PVI	Brine	XG	SCL
2.00			
4.00			

Table A2. Images segmented of the evaluation of the silica NPs in the polymer flooding in the function of the porous volume injected (PVI) during the recovery agent injection. This part shows the reference nanofluid (NF) and the xanthan gum with NPs (XG_NP). Due to different injection problems, the Scleroglucan with NPs was unsuccessful.

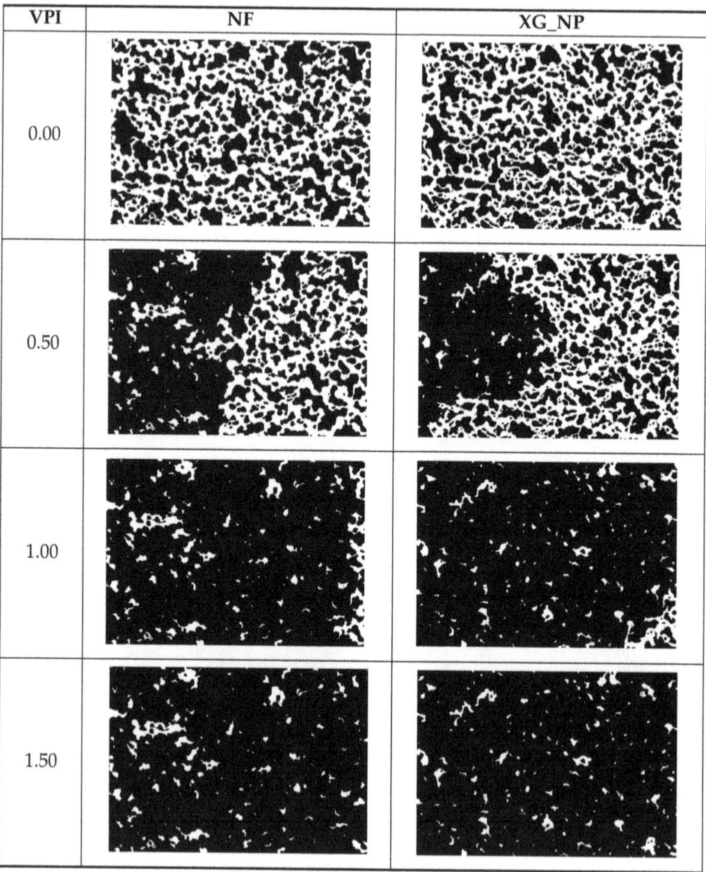

VPI	NF	XG_NP
0.00		
0.50		
1.00		
1.50		

Table A3. Images segmented from the microfluidic tests to evaluate the effect of the flow rate on the recovery performance of the xanthan gum with NPs (XG_NP) in comparison with the reference nanofluid (NF).

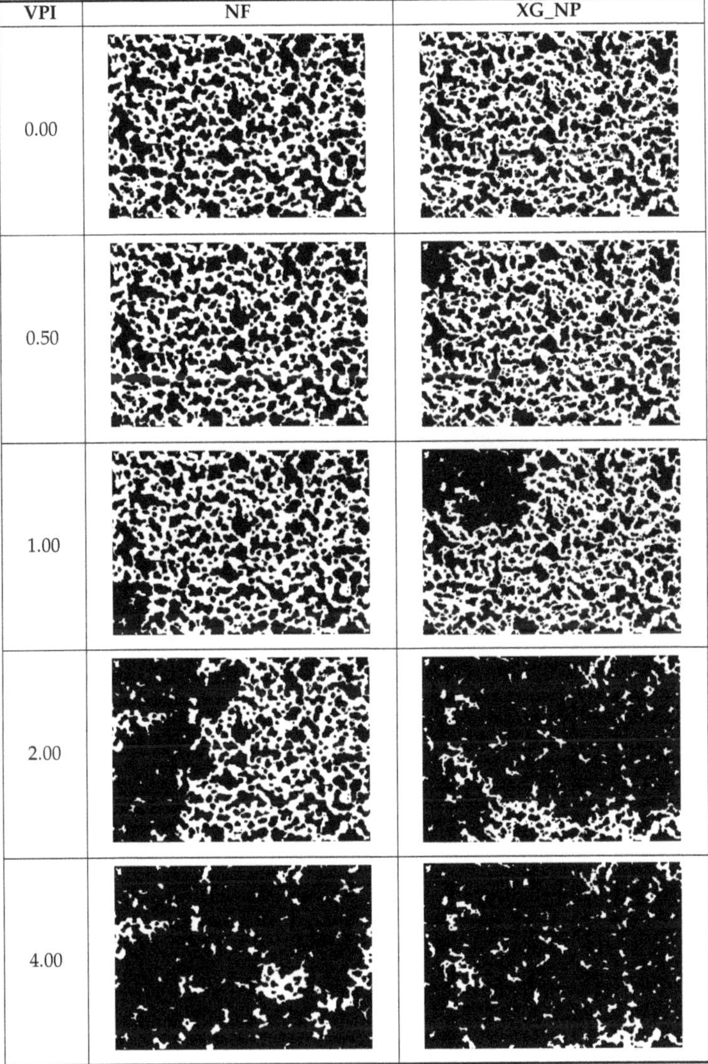

References

1. Lake, L.W. *Enhanced Oil Recovery*; Print-on-demand.; Univ. Co-op.: Austin, TX, USA, 2011; ISBN 978-0-8400-6603-9.
2. Lyons, W.C. Enhanced oil recovery methods. In *Working Guide to Reservoir Engineering*; Elsevier: Burlington, MA, USA, 2010; pp. 279–310.
3. Speight, J.G. Nonthermal methods of recovery. In *Introduction to Enhanced Recovery Methods for Heavy Oil and Tar Sands*; Elsevier: Laramie, WY, USA, 2016; pp. 353–403.
4. Sorbie, K.S.; Sorbie, K.S. Structure of the main polymers used in improved oil recovery (IOR). In *Polymer-Improved Oil Recovery*; Springer: Dordrecht, The Netherlands, 1991; pp. 6–36.

5. Park, J.K.; Khan, T. Other microbial polysaccharides: Pullulan, scleroglucan, elsinan, levan, alternant, dextran. In *Handbook of Hydrocolloids*, 2nd ed.; Elsevier Inc.: Cambridge, UK, 2009; pp. 592–614, ISBN 9781845695873.

6. Zhou, W.; Zhang, J.; Han, M.; Xiang, W.; Feng, G.; Jiang, W. Application of hydrophobically associating water-soluble polymer for polymer flooding in china offshore heavy oilfield. In Proceedings of the International Petroleum Technology Conference, Dubai, UAE, 4–6 December 2007.

7. Bluhm, T.L.; Deslandes, Y.; Marchessault, R.H.; Pérez, S.; Rinaudo, M. Solid-state and solution conformation of scleroglucan. *Carbohydr. Res.* **1982**, *100*, 117–130. [CrossRef]

8. Sletmoen, M.; Stokke, B.T. Higher order structure of (1,3)-β-D-glucans and its influence on their biological activities and complexation abilities. *Biopolym.* **2008**, *89*, 310–321. [CrossRef]

9. Viñarta, S.C.; Delgado, O.D.; Figueroa, L.I.C.; Fariña, J.I. Effects of thermal, alkaline and ultrasonic treatments on scleroglucan stability and flow behavior. *Carbohydr. Polym.* **2013**, *94*, 496–504. [CrossRef]

10. Perry, B. Performance history on use of biopolymer in springer sand waterflood in southern oklahoma. In Proceedings of the Fall Meeting of the Society of Petroleum Engineers of AIME, San Antonio, TX, USA, 8–11 October 1972.

11. Littmann, W.; Kleinitz, W.; Christensen, B.E.; Stokke, B.T.; Haugvallstad, T. Late Results of a polymer pilot test: Performance, simulation adsorption, and xanthan stability in the reservoir. In Proceedings of the SPE/DOE Enhanced Oil Recovery Symposium, Tulsa, OK, USA, 22–24 April 1992.

12. Skauge, T.; Spildo, K.; Skauge, A. Nano-sized particles for EOR. In Proceedings of the SPE Improved Oil Recovery Symposium, Tulsa, OK, USA, 24–28 April 2010.

13. Hendraningrat, L.; Li, S.; Torsæter, O. A coreflood investigation of nanofluid enhanced oil recovery in low-medium permeability Berea sandstone. In Proceedings of the SPE International Symposium on Oilfield Chemistry, The Woodlands, TX, USA, 8–10 April 2013; Volume 2, pp. 728–741.

14. Bila, A.; Stensen, J.Å.; Torsæter, O. Experimental Investigation of Polymer-Coated Silica Nanoparticles for Enhanced Oil Recovery. *Nanomaterials* **2019**, *9*, 822. [CrossRef]

15. Kazemzadeh, Y.; Shojaei, S.; Riazi, M.; Sharifi, M. Review on application of nanoparticles for EOR purposes: A critical review of the opportunities and challenges. *Chin. J. Chem. Eng.* **2019**, *27*, 237–246. [CrossRef]

16. Chengara, A.; Nikolov, A.D.; Wasan, D.T.; Trokhymchuk, A.; Henderson, D. Spreading of nanofluids driven by the structural disjoining pressure gradient. *J. Colloid Interface Sci.* **2004**, *280*, 192–201. [CrossRef] [PubMed]

17. El-Diasty, A.I.; Aly, A.M. Understanding the mechanism of nanoparticles applications in enhanced oil recovery. In Proceedings of the SPE North Africa Technical Conference and Exhibition, Cairo, Egypt, 14–16 September 2015.

18. Afrapoli, M.S. Experimental study of displacement mechanisms in microbial improved oil recovery processes. Ph.D. Thesis, Norwegian University of Science and Technology (NTNU), Trondheim, Norway, 2010.

19. Suleimanov, B.A.; Ismailov, F.S.; Veliyev, E.F. Nanofluid for enhanced oil recovery. *J. Pet. Sci. Eng.* **2011**, *78*, 431–437. [CrossRef]

20. Maghzi, A.; Mohebbi, A.; Kharrat, R.; Ghazanfari, M.H. Pore-scale monitoring of wettability alteration by silica nanoparticles during polymer flooding to heavy oil in a five-spot glass micromodel. *Transport. Porous Media* **2011**, *87*, 653–664. [CrossRef]

21. Maghzi, A.; Kharrat, R.; Mohebbi, A.; Ghazanfari, M.H. The impact of silica nanoparticles on the performance of polymer solution in presence of salts in polymer flooding for heavy oil recovery. *Fuel* **2014**, *123*, 123–132. [CrossRef]

22. Cheraghian, G.; Khalilinezhad, S.S. Effect of nanoclay on heavy oil recovery during polymer flooding. *Pet. Sci. Technol.* **2015**, *33*, 999–1007. [CrossRef]

23. Khalilinezhad, S.S.; Cheraghian, G.; Karambeigi, M.S.; Tabatabaee, H.; Roayaei, E. Characterizing the role of clay and silica nanoparticles in enhanced heavy oil recovery during polymer flooding. *Arab. J. Sci. Eng.* **2016**, *41*, 2731–2750. [CrossRef]

24. Saha, R.; Uppaluri, R.V.S.; Tiwari, P. Silica nanoparticle assisted polymer flooding of heavy crude oil: Emulsification, rheology, and wettability alteration characteristics. *Ind. Eng. Chem. Res.* **2018**, *57*, 6364–6376. [CrossRef]

25. Corredor, L.; Maini, B.; Husein, M. Improving polymer flooding by addition of surface modified nanoparticles. In Proceedings of the SPE Asia Pacific Oil and Gas Conference and Exhibition, Brisbane, Australia, 23–25 October 2018.

26. Mohammadi, M.; Khorrami, M.K.; Ghasemzadeh, H. ATR-FTIR spectroscopy and chemometric techniques for determination of polymer solution viscosity in the presence of SiO_2 nanoparticle and salinity. *Spectrochim. Acta-Part A* **2019**, *220*, 117049. [CrossRef] [PubMed]

27. Kennedy, J.R.M.; Kent, K.E.; Brown, J.R. Rheology of dispersions of xanthan gum, locust bean gum and mixed biopolymer gel with silicon dioxide nanoparticles. *Mater. Sci. Eng. C* **2015**, *48*, 347–353. [CrossRef] [PubMed]

28. Khishvand, M.; Akbarabadi, M.; Piri, M. Micro-scale experimental investigation of the effect of flow rate on trapping in sandstone and carbonate rock samples. *Adv. Water Resour.* **2016**, *94*, 379–399. [CrossRef]

29. Zhang, H.; Ramakrishnan, T.S.; Nikolov, A.; Wasan, D. Enhanced oil recovery driven by nanofilm structural disjoining pressure: Flooding experiments and microvisualization. *Energy Fuels* **2016**, *30*, 2771–2779. [CrossRef]

30. Civan, F. Injectivity of the water-flooding wells. In *Reservoir Formation Damage*; Elsevier: Waltham, MA, USA, 2015; pp. 343–377.

31. Daripa, P.; Dutta, S. Modeling and simulation of surfactant–polymer flooding using a new hybrid method. *J. Comput. Phys.* **2017**, *335*, 249–282. [CrossRef]

32. Ferreira, V.H.; Moreno, R.B. Workflow for oil recovery design by polymer flooding. In Proceedings of the International Conference on Ocean, Offshore and Arctic Engineering, Madrid, Spain, 17–22 June 2018.

33. Pradhan, S.; Shaik, I.; Lagraauw, R.; Bikkina, P. A semi-experimental procedure for the estimation of permeability of microfluidic pore network. *MethodsX* **2019**, *6*, 704–713. [CrossRef] [PubMed]

34. Toriwaki, J.; Yoshida, H. *Fundamentals of Three-Dimensional Digital Image Processing*; Springer: London, UK, 2009; ISBN 9781849967440.

 *nanomaterials*

Article

A Core Flood and Microfluidics Investigation of Nanocellulose as a Chemical Additive to Water Flooding for EOR

Reidun C. Aadland [1,*] , Salem Akarri [1], Ellinor B. Heggset [2] , Kristin Syverud [2,3] and Ole Torsæter [1]

1   PoreLab Center of Excellence, Department of Geoscience and Petroleum, Norwegian University of Science and Technology (NTNU), N-7491 Trondheim, Norway; salem.s.f.akarri@ntnu.no (S.A.); ole.torsater@ntnu.no (O.T.)
2   RISE PFI, N-7491 Trondheim, Norway; ellinor.heggset@rise-pfi.no (E.B.H.); kristin.syverud@rise-pfi.no (K.S.)
3   Department of Chemical Engineering, Norwegian University of Science and Technology (NTNU), N-7491 Trondheim, Norway
*   Correspondence: reidun.aadland@ntnu.no

Received: 6 June 2020; Accepted: 30 June 2020; Published: 1 July 2020

Abstract: Cellulose nanocrystals (CNCs) and 2,2,6,6-tetramethylpiperidine-1-oxyl (TEMPO)-oxidized cellulose nanofibrils (T-CNFs) were tested as enhanced oil recovery (EOR) agents through core floods and microfluidic experiments. Both particles were mixed with low salinity water (LSW). The core floods were grouped into three parts based on the research objectives. In Part 1, secondary core flood using CNCs was compared to regular water flooding at fixed conditions, by reusing the same core plug to maintain the same pore structure. CNCs produced 5.8% of original oil in place (OOIP) more oil than LSW. For Part 2, the effect of injection scheme, temperature, and rock wettability was investigated using CNCs. The same trend was observed for the secondary floods, with CNCs performing better than their parallel experiment using LSW. Furthermore, the particles seemed to perform better under mixed-wet conditions. Additional oil (2.9–15.7% of OOIP) was produced when CNCs were injected as a tertiary EOR agent, with more incremental oil produced at high temperature. In the final part, the effect of particle type was studied. T-CNFs produced significantly more oil compared to CNCs. However, the injection of T-CNF particles resulted in a steep increase in pressure, which never stabilized. Furthermore, a filter cake was observed at the core face after the experiment was completed. Microfluidic experiments showed that both T-CNF and CNC nanofluids led to a better sweep efficiency compared to low salinity water flooding. T-CNF particles showed the ability to enhance the oil recovery by breaking up events and reducing the trapping efficiency of the porous medium. A higher flow rate resulted in lower oil recovery factors and higher remaining oil connectivity. Contact angle and interfacial tension measurements were conducted to understand the oil recovery mechanisms. CNCs altered the interfacial tension the most, while T-CNFs had the largest effect on the contact angle. However, the changes were not significant enough for them to be considered primary EOR mechanisms.

Keywords: enhanced oil recovery; chemical flooding; nanocellulose; cellulose nanocrystals; TEMPO-oxidized cellulose nanofibrils; microfluidics

---

1. Introduction

The majority of existing oil fields are in their tail-end production, where most of the easily accessible oil has already been produced. The remaining oil is difficult to recover, and there is a low proportion of new fields for exploration, as most of the basins that might contain oil have already been explored, and many of the unexplored basins lie in remote and environmentally sensitive areas of the world (e.g., the Arctic) [1]. Therefore, it is important to try to extend the lifetime of already operating fields. For example, the average recovery rate for the oil fields on the Norwegian continental shelf is approximately 50%, thus, resulting in a high amount of unrecovered oil that is not producible with the current technology [2]. The application of enhanced oil recovery (EOR) techniques entails methods that could improve the oil recovery. Research and technological development have been directed to advance the techniques of enhanced oil recovery. Recently, nanocelluloses have been introduced as environmentally friendly nanoparticles for EOR applications [3–7].

Cellulose is an abundant biopolymer derived from various sources, usually from wood. A tree produces 13–14 g of cellulose per day, and the total production of cellulose all over the world is estimated to be 7.5×10^{10} tons per year [8,9]. In the cell wall of plants, cellulose molecules are formed as solid structures made from bundles of cellulose molecules held together by inter- and intramolecular hydrogen bonds, with dimensions in the nanoscale. Today, it is possible to extract these nanoscaled structures from plants by various methods. Depending on the production strategy, different structures of nanocellulose can be formed. By subjecting cellulose to controlled acid hydrolysis, cellulose nanocrystals (CNCs) can be created, while cellulose nanofibrils (CNFs) can be obtained by using high shear forces, e.g., a high-pressure homogenizer [10], often after a chemical pretreatment has been applied. CNCs from wood are usually 3–5 nm wide, and have lengths ranging from 100–200 nm [11], while CNFs usually have a diameter in the range of 5–60 nm and lengths of several micrometers [10]. The OSPAR Commission reported that CNCs and CNFs pose little or no risk to the offshore environment [12].

Thermal stability is an important attribute when testing new particles for EOR purposes, as temperatures in reservoirs can get quite high (>90 °C) [13]. Heggset et al. [14] studied the temperature stability of CNCs and CNFs, where it was found that both types of particles exhibited superior temperature stability when compared to e.g., the biopolymer xanthan. Furthermore, in another study using modified nanocellulose, it was found that when nanocellulose was subjected to elevated temperatures it experienced a slower loss in viscosity compared to the synthetic polymer, hydrolyzed polyacrylamide (HPAM) [15]. Therefore, CNCs and CNFs might serve as a sustainable and effective alternative EOR technique for offshore oil fields.

Molnes et al. [6] did core-flood experiments using CNCs and observed a slight increase in oil recovery (3.4%) when the nanofluid was injected as a tertiary recovery agent. Aadland et al. [3] did a high-temperature core flood with CNCs and showed that the nanoparticles contributed to a slight increase in incremental oil recovery (1.2%). Kusanagi et al. [5] did tertiary core flood experiments using 2,2,6,6-tetramethylpiperidine-1-ox (TEMPO)-oxidized cellulose nanofibers (T-CNFs) and found that additional oil was produced (8.6%). However, they observed filtration in the porous media and poor injectivity. Experiments using surface-functionalized nanocellulose showed that the particles contributed to a 3–17% increase in oil recovery when injected after water flooding. From microfluidic experiments using the same particles, three main oil recovery mechanisms were discovered: emulsification, dragging, and wettability alteration [16]. Emulsification was also one of the main findings from a previous study Wei et al. [15] conducted.

A recovery agent, i.e., nanofluid or low salinity brine, can be injected in a secondary or tertiary mode. Secondary mode corresponds to the injection at initial water saturation and mainly displaces a large connected body of oil in the porous medium, while the tertiary mode is the injection after reaching residual oil saturation via the secondary-mode fluid, aiming to mobilize the remaining trapped oil clusters. Therefore, the response to injecting a recovery agent differs according to the mode, as shown in several studies [17–22]. In addition, the response is also significantly controlled by the wettability condition of the rock, as it affects the initial distribution of fluids within the porous medium, as well as the displacement dynamics [23–28]. Several studies tested recovery agents at different wettability conditions, i.e., water wet and intermediate wet, however, the reported results were not consistent [19–21,29,30]. A study suggested that the effect of aging, in terms of oil recovery, was dependent on the oil-brine-rock system, where, for example, rocks of the same type with different clay content would show different responses. As previously mentioned, the temperature is another important parameter in relation to oil recovery, which also affects the recovery response. Therefore, it is vital to establish an understanding of how new EOR candidates behave in different recovery modes, wettability conditions, and temperatures.

Mobilization of the residual oil in a porous medium is considerably governed by the capillary number (N_C) [31]. It is defined in Equation (1) [32]; where q_w is the interstitial velocity (m/s), μ_w is the viscosity (Pa s) and σ_{ow} is the interfacial tension (N/m) between oil and water. The flow in a porous medium is dominated by viscous forces if the capillary number is high, and capillary forces if the capillary number is low. A higher capillary number can be achieved by decreasing the interfacial tension, or by increasing the viscosity or the velocity of the displacing phase. The capillary number at the end of a water flood usually ranges from 10^{-6} to 10^{-4} [33].

$$N_C = \frac{q_w \times \mu_w}{\sigma_{ow}} \tag{1}$$

The capillary number is directly related to the amount of trapped oil in a porous medium.

For brine flooding, studies [34,35] have shown that high capillary numbers (achieved by increasing the injection rate) induced lower volumes of trapped oil in natural rocks and glass-bead packs. However, for nanofluids a different trend was observed, where high capillary numbers yielded lower oil recovery factors. The low oil recovery was explained by that nanoparticles require sufficient time for altering wettability via structural-disjoining pressure, or that they agglomerate at high injection rates [35,36].

Microfluidic micromodels have been a vital technique in EOR applications, since they provide micro-visualization of the fluid flow behavior, which can be recorded for qualitative observation, quantitative analysis, and simulation studies. Microfluidic micromodels have been used for assessing surfactant-polymer flooding mixed with nanoclay for improving heavy oil recovery [37], evaluating polymer EOR for unconsolidated sand reservoirs [38], investigating EOR mechanisms associated with injecting silica nanoparticles [39], and the screening of surface-modified silica nanoparticles for EOR [40]. They also have been used to study the effect of the polymer concentration and injection rate on the sweep efficiency [41], examining the impact of fluid rheology on oil recovery [42], and simulation of fluid configurations captured from imbibition and drainage experiments on a micromodel [43].

In addition to the recovery factor, the acquired images from microfluidic studies can be processed to evaluate the cluster size distribution and connectivity of the non-wetting phase. Cluster size distribution is affected by wettability condition [44], injection rate [34], and interfacial tension [45]. The connectivity of the non-wetting phase in a porous medium can be described by the Euler characteristic/number (E) [46,47]. For 2D images, the Euler number is given by Equation (2), where C is the number of isolated components in the image, and H is the number of holes within the components [47,48]. The aforementioned factors affecting the cluster size distribution might influence the connectivity, since Equation (2) is dependent on the number of clusters in the image.

$$E = C - H \tag{2}$$

In the current experimental study, core floodings and microfluidic experiments have been conducted, to further investigate the potential of nanocelluloses for enhanced oil recovery. On the core-scale, the main objective was to perform a comprehensive study on CNCs as recovery agent in a secondary and tertiary mode, in water-wet and intermediate-wet systems, and at high and low temperatures. In addition, a novel approach was tested for one core plug, where the idea was to reuse the same natural pore structure (same core plug) after wettability restoration, excluding the pore architecture effect on trapping efficacy. Furthermore, T-CNFs, which are of different structure and size compared to CNCs, were evaluated on core-scale to see their ability in the mobilization of trapped oil, compared to CNCs at the same conditions. Moreover, 2D glass microchips of the same pore-structure and wettability were used to obtain micro-scale comparisons between the injection of brine, CNCs, and T-CNFs. The studied micro-scale parameters were dynamic changes in the oil connectivity, oil recovery, and residual cluster size distribution. The microchips were also used to show the effect of a higher flow rate on oil recovery, size distribution, and oil connectivity. This study aims to contribute towards filling the knowledge gap in the nanocellulose literature for EOR applications.

2. Materials

2.1. Rock

The core plugs used in this study were extracted from a Berea sandstone block and had an average length and diameter of 10 cm and 3.8 cm, respectively. The block was acquired from a quarry in Ohio, USA and was purchased from Berea Sandstone Petroleum Cores (Berea Sandstone Petroleum Cores, Vermilion, OH, USA). The core plugs were rinsed in a Soxhlet apparatus with methanol and dried in an oven at 60 °C, prior to the core floods. Permeability and porosity measurements were performed on the dry core plugs. Core properties are listed in Table 1.

Table 1. Properties of the cores used in experiments.

Core	Length	Diameter	Pore Volume	Permeability	Porosity
(no.)	(cm)	(cm)	(mL)	(mD)	(%)
1	10.0	3.8	19.3	781	17.5
2	10.0	3.7	18.3	896	15.1
3	9.9	3.8	17.5	883	15.9
4	9.7	3.8	18.4	1111	17.1
5	9.9	3.8	18.5	1067	16.9
6	9.9	3.8	18.8	928	17.2
7	9.9	3.8	18.8	768	17.2
8	9.9	3.8	18.1	771	16.5
9	9.9	3.8	18.3	727	16.8
10	9.9	3.8	17.5	832	16.0

2.2. Microfluidic Chip

Five borosilicate glass microfluidic chips of the same pore-network structure were used in this study (Micronit Microfluidics, Enschede The Netherlands). The dimensions of the chip are 45 × 15 × 1.8 mm. The chips contain a porous medium (20 × 10 × 0.02 mm) with a pore-network structure representing actual rock-pore structures. The pore volume, permeability, and porosity are 2.3 μL, 2.5 D, and 57%, respectively.

2.3. Brine

All experiments were performed using low salinity water (LSW), which consisted of 0.1 wt.% sodium chloride (NaCl) prepared from NaCl (Sigma-Aldrich, St. Louis, MO, USA) and de-ionized water (DIW). Properties are listed in Table 2.

2.4. Nanocelluloses

2.4.1. Cellulose Nanocrystals

The cellulose nanocrystals (CNCs) were acquired from the University of Maine. The material was manufactured at the Forest Products Laboratory in Madison, USDA (U.S. Dep. of Agriculture, USA). The CNCs were produced by acid hydrolysis of softwood pulp, where 64% (by mass) sulphuric acid was used to hydrolyze amorphous regions of the cellulose material, yielding acid-resistant crystals [49]. The stock-dispersion had a concentration of 12 wt.%. Properties are listed in Tables 2 and 3, and an atomic force microscopy (AFM) image of the particles can be seen in Figure 1. In the experiments, the nanofluid was diluted to 1.0 wt.% CNCs using LSW.

Table 2. Fluid properties.

Fluid	Density (g/cm^3)		Viscosity (cP)	
	24 °C	60 °C	24 °C	60 °C
0.1 wt.% NaCl	1.00	0.98	0.91	0.47
1 wt.% CNCs in 0.1 wt.% NaCl	1.00	1.01	1.40	1.09
0.5 wt.% CNCs in 0.1 wt.% NaCl	0.99	1.00	1.33	1.24
0.1 wt.% T-CNFs in 0.1 wt.% NaCl	1.01	-	3.67	-
Crude oil C	0.91	0.89	55.90	12.19
Crude oil D	0.89	0.87	20.74	5.88

2.4.2. TEMPO-Oxidized Cellulose Nanofibrils

TEMPO-oxidized cellulose nanofibrils (T-CNFs) were produced at RISE PFI (Trondheim, Norway). For the production of T-CNFs, never-dried, bleached softwood pulp fibers were used as the source material. The preparation was performed using 2,2,6,6-tetramethylpiperidine-1-oxyl (TEMPO) radical-mediated oxidation, as previously described by Isogai et al. [50]. The TEMPO-oxidized pulp was afterwards pretreated in a Mazuko-grinder before further fibrillation. The fibrillation was done using a Rannie 15 type 12.56× homogenizer (APV, SPX Flow Technology, Silkeborg, Denmark), and the samples were fibrillated for four passes with a pressure drop of 1000 bar in each pass. Carboxylate content was determined using conductometric titration as previously described [51,52]. The equipment used was a 902 Titrando (Methrom AG, Herisau, Switzerland), an 856 conductivity module and Tiamo software (Metrohm AG, Herisau, Switzerland). The stock dispersion of T-CNFs had a concentration of 0.66 wt.% and was diluted to 0.1 wt.% using LSW. Properties are listed in Tables 2 and 3, and an AFM image of the particles can be seen in Figure 1.

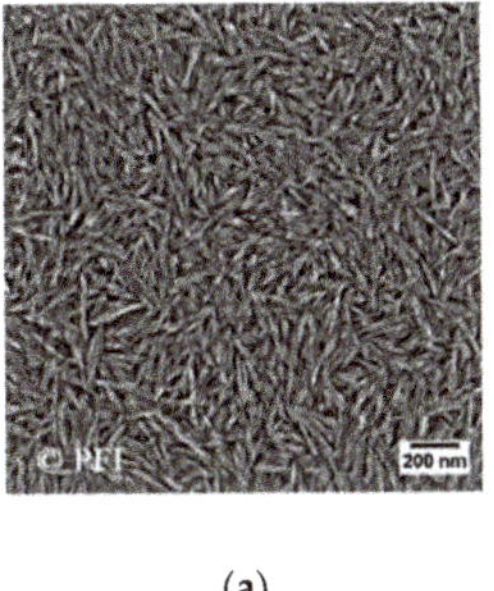

(**a**)

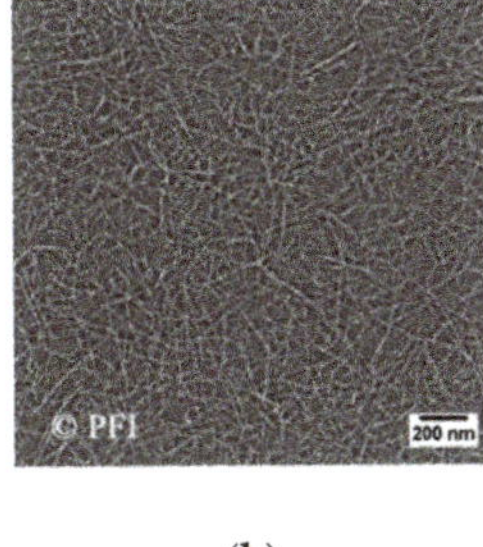

(**b**)

Figure 1. Atomic force microscopy (AFM) images of cellulose nanocrystals (**a**) and 2,2,6,6-tetramethylpiperidine-1-oxyl (TEMPO)-oxidized cellulose nanofibrils (**b**). From the images, it can be seen that both cellulose nanocrystals (CNCs) and TEMPO-oxidized cellulose nanofibrils (T-CNFs) consists of elongated particles, and that T-CNFs are longer compared to the CNCs. They both exhibit anisotropic properties. The procedure for how the AFM images were obtained is explained in Aadland [53].

From the particle size measurement with dynamic light scattering (DLS), it was observed that CNCs are, to a greater extent, a monodisperse sample, while T-CNF was polydisperse. DLS is not considered a good technique to determine the size of these rod-shaped particles, as it is focused towards spherical particles. Nevertheless, it gives a relative particle size and allows us to compare the particles against each other. From the obtained AFM image (Figure 1b) it can be seen that the diameter for T-CNF is approx. 15 nm, which is within the range that has been reported before in the literature for cellulose nanofibrils. Cellulose nanofibrils usually have a diameter in the range of 5–60 nm, and can be several micrometers in length [10].

Table 3. Charge density and properties of the nanocellulose qualities.

Sample	Charge Density * (mmol/g)	Functional Groups in Significant Amounts	Zeta Potential	Apparent Size by DLS (nm)
CNCs	approx. 0.3 **	–OH, –SO$_3$H	−40.1 ± 2.5	123 ± 0–164 ± 2 nm
T-CNFs	1.13	–OH, –COOH, –CHO	−41.7 ± 2.2	1019 ± 297 nm

* This is carboxylic acids for T-CNFs and sulphate half ester for CNCs. ** Determined by ICP-AA. The procedure for zeta potential and apparent size by dynamic light scattering (DLS) can be found in Aadland [54].

The obtained characteristics for the CNCs are in accordance with result from the previous literature [7,55,56].

2.5. Oil

Two types of crude oil were used during the experiments, denoted crude oil C and crude oil D. Both were taken from the same field in the Norwegian Sea, but they exhibit different properties (Table 3). The oils were filtered twice through a five μm Millipore filter under vacuum to remove impurities. A saturates, aromatics, resins, asphaltenes (SARA) analysis was performed for both the oils, and can be found in Table 4. From the analysis, it was seen that crude oil D contained much fewer resins and asphaltenes compared to crude oil C. Resins and asphaltenes are polar compounds, which are considered the surface-active components in crude oil [57].

Table 4. Saturates, aromatics, resins, asphaltenes (SARA) analysis of crude oil.

Type of Oil	Weight Percent (Normalized)			
	Saturates	Aromatics	Resins	Asphaltenes
Crude oil C	66.21	25.78	7.69	0.32
Crude oil D	71.57	20.81	7.44	0.18

3. Experimental Methods

3.1. Fluid Interaction Measurements

3.1.1. Fluid-Fluid Interactions

Interfacial tension (IFT) was measured using the drop shape analyzer DSA 100S with acquisition software (Kruss GmbH, Hamburg, Germany). The IFT was automatically calculated in the software ADVANCE (Kruss GmbH, Hamburg, Germany), by applying the pendant drop technique, where the contour of the drop was determined with image analysis. Measurements were taken every five minutes for 12 h. The experiment was conducted at ambient temperature and pressure.

3.1.2. Fluid-Solid Interactions

The contact angle was measured using the same apparatus and acquisition software that was applied for IFT. The measurement cell was filled with the respective fluid (i.e., brine or nanocellulose) and an oil droplet was placed underneath a glass substrate. The static contact angle was measured using the captive bubble method with the software ADVANCE. The contact angle was measured at 5 min intervals for 12 h and was performed at ambient conditions.

3.2. Core Flood Study

3.2.1. Experimental Setup

The experimental setup of the core floods can be seen in Figure 2. A high-performance liquid chromatography (HPLC) pump (Teledyne ISCO, Lincoln, NE, USA) was used, with Exxsol™ D60 (ExxonMobil Chemical Europe, Machelen, Belgium) as pump fluid. Three cylinders were installed inside the heating cabinet, containing the relevant fluids-crude oil, brine, and nanofluid, respectively. For crude oil and nanofluid, a piston was installed inside the cylinder to prevent mixing with Exxsol™ D60. Prior to each flooding stage, the respective fluid was flooded through the bypass line and out to the effluent collector to reduce the dead volume. The back-pressure valve was only installed and applied during the high-temperature floods.

The core was installed horizontally in a Hassler core cell and sleeve a pressure within 20–25 bar was maintained. The pressure was logged across the core throughout the experiment, and effluent samples were collected every one-fourth pore volume (PV) during water- and nano flooding.

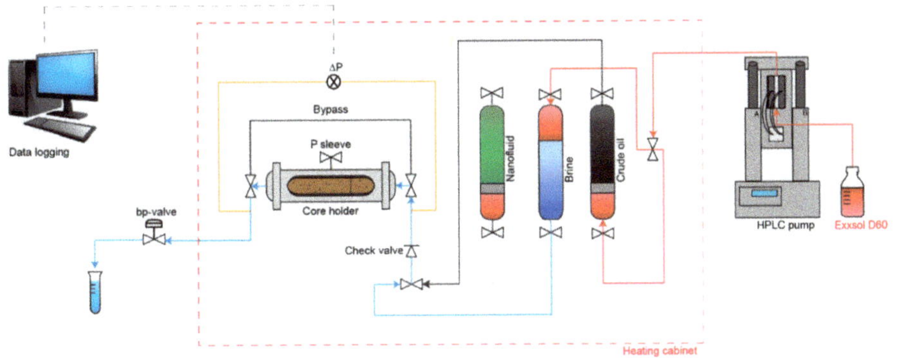

Figure 2. Schematic illustration of the experimental setup of the core flood.

3.2.2. Core Flood Experiments

Ten core flood experiments were conducted, where the overall goal was to determine if nanocelluloses have potential as enhanced oil recovery agents. The experiments were designed to test the performance of the fluid under various conditions. The injection mode, type of oil, temperature, wettability of rock, and particle type were the variables investigated. The experiments were classified into three parts based upon their research objectives (see Table 5 for details).

The first part compares the induced oil recovery by CNCs to brine at fixed conditions, including the pore structure via reusing the same core plug. The second part aims at exploring the response of CNCs as a recovery agent in a secondary and tertiary mode, in water-wet and intermediate-wet systems, and at high and low temperatures. The third part compares T-CNFs, to CNCs in terms of the ability to mobilize trapped oil under the same conditions.

Table 5. Overview of the different tests conducted. The letters A and B denote that the experiments were conducted on the same core plug. The test number corresponds to the core number in Table 1.

Part	Test No.	Fluids			Conditions	
		Secondary Agent	Tertiary Agent	Oleic Phase	Temp. (°C)	Aging Time (Weeks)
1	1A	0.1 wt.% NaCl	-			-
	1B	1.0 wt.% CNCs	-	Crude oil C		-
2	2	1.0 wt.% CNCs	-		24	-
	3	0.1 wt.% NaCl	1.0 wt.% CNCs			-
	4	0.1 wt.% NaCl	1.0 wt.% CNCs			5
	5	1.0 wt.% CNCs	-			5
	6	0.1 wt.% NaCl	1.0 wt.% CNCs	Crude oil D		-
	7	1.0 wt.% CNCs	-		60	-
	8	0.1 wt.% NaCl	1.0 wt.% CNCs			5
	9	1.0 wt.% CNCs	-			7
3	10	0.1 wt.% NaCl	0.1 wt.% T-CNF		24	-

3.2.3. Core Flood Procedure

A 100% brine saturated core was installed in the core holder. Initial saturations were established by drainage using crude oil injected at two different flow rates (1.0 and 10 mL/min). The oil recovery experiment was conducted either as a secondary-or tertiary recovery technique, where the respective fluid was injected using two flow rates (0.3 and 3.0 mL/min). Each flow rate continued until pressure stabilized and no more oil was produced. For secondary mode recovery, LSW or nanofluid was injected after establishing initial conditions in the core. Tertiary mode recovery consisted of injecting LSW as secondary mode, then injecting the nanofluid. After each nanoflood stage (secondary or tertiary) a post flush with LSW was injected at 3.0 mL/min, to see if any incremental oil could be produced as a result of creating a new flow pattern in the core induced by the nanoparticles.

3.2.4. Aging

Four cores were aged using crude oil D (Test 4, 5, 8, and 9). Aging was done to make the cores more intermediate wet. First, initial saturations were established in the core using the same procedure as explained in the previous section. Afterwards, the core was placed in a container and immersed in crude oil D. The container was stored in an oven at 60 °C. Three of the cores were aged for five weeks (35 days), while one core was aged for seven weeks (49 days).

3.3. Microfluidic Study

3.3.1. Experimental Setup

The experimental setup of the microfluidic experiment is presented in Figure 3. A Micronit EOR platform was used to conduct the microfluidic experiments under a microscope (Olympus SZX7, Olympus, Langhus, Norway) integrated with a digital camera (Olympus UC90, Olympus, Langhus, Norway). Olympus Stream Basic 2.1 (Olympus, Langhus, Norway) was used for image acquisition, as well as experiment monitoring, via a live feed. A syringe pump (Harvard Apparatus Pump 33 DDS, Holliston, MA., USA) and CODAN luer-lock syringes were used for the fluid injection. The syringe was connected to a 4-way valve, which was used to bleed the line from the syringe when changing to a new fluid. The production line was connected to a three-way valve directing the fluids to a waste beaker during the experiment or allowing the vacuuming of the system using Edwards RV3 vacuum pump (Edwards, Lørenskog, Norway).

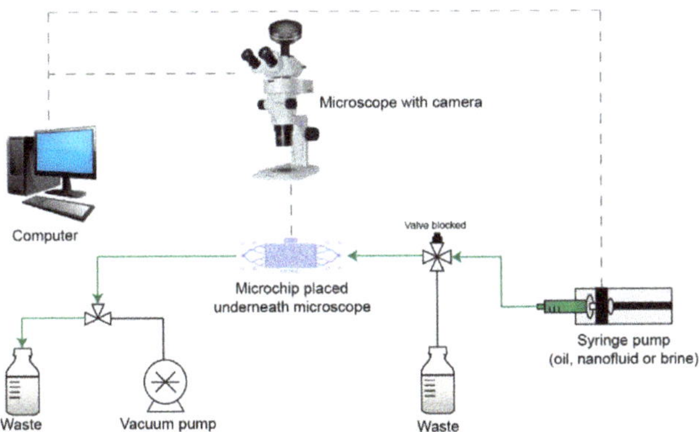

Figure 3. Schematic illustration of the experimental setup of the microfluidic experiment.

3.3.2. Microfluidics Experiments

Five microfluid experiments were conducted at room temperature, which are classified into two groups: high injection-rate and low-injection rate experiments. They were designed to evaluate the dynamic change in oil recovery, connectivity, and clustering, enabling a comparison between the three injected fluids during the low rate injection scheme. In addition, it aims at evaluating the performance of the nanofluids when they are injected with a high rate. Table 6 presents the details of the injection and image acquisition. The analyzed area, which is with the same pore structure for the five experiments, was selected in the middle of the microchip to avoid capillary end effects.

Table 6. Details of injection and image acquisition for microfluidics experiments.

Test No.	Recovery Agent	Flow Rate (μL/min)	Capillary Number	Duration (min)	Pore Volume Injected	Analyzed Area (mm^2)	Image Number
M1	0.1 wt.% NaCl		1.2×10^{-6}				
M2	1.0 wt.% CNCs	0.18	2.1×10^{-6}	23.0	1.8		31
M3	0.1 wt.% T-CNFs		5.0×10^{-6}			28.1	
M4	1.0 wt.% CNCs	1.80	2.1×10^{-5}	10.2	8.0		25
M5	0.1 wt.% T-CNFs		5.0×10^{-5}				

3.3.3. Microfluidic Procedure

After inserting a clean and dry microfluidic chip in the flooding platform, the system was vacuumed to remove the air through the production line. After that, brine (0.1 wt.% NaCl) was injected to fully saturate the chip at a constant rate of 0.1 mL/min. It was followed by injecting the non-wetting phase at three different flow rates (0.006, 0.06, and 0.5 mL/min) to establish initial oil and irreducible water saturations in the microchip. After acquiring an image of the initial state, a secondary recovery was performed. During the recovery stage, images were taken to capture the dynamic changes.

3.3.4. Image Processing and Analysis

The acquired images were processed and analyzed using Fiji [58] with an installation of BoneJ [59] plugin. The images were firstly segmented using a color thresholding method to extract the non-wetting phase. The resulted binary images were utilized for cluster analysis, in order to calculate the total area of the oil (A_o), cluster size distribution, number of clusters (C), number of holes (H), and size of clusters. Figure 4 illustrates an image of a saturated microchip through the process. Based on these parameters, the change in oil recovery, oil connectivity, and cluster number were evaluated as a function of pore volume injected (PVI). The oil recovery is given by Equation (3). The oil connectivity is described by the Euler number, see Equation (2). In this work, the Euler number is normalized by the Euler number of the oil at the initial stage i.e., pre-imbibition. A normalized Euler number (X) of 1 represents the pre-flooding stage, which is the maximum value and state of connectivity, and the more disconnected the oil becomes, the farther the value is from 1. For example, the non-wetting phase with a connectivity of X = 0.2 is more disconnected than X = 0.4. In addition, the normalized cumulative residual oil volume as a function of oil cluster size was obtained.

$$Oil\ recovery = \frac{A_{oPVI=0} - A_{oPVI\geq0}}{A_{oPVI=0}} \tag{3}$$

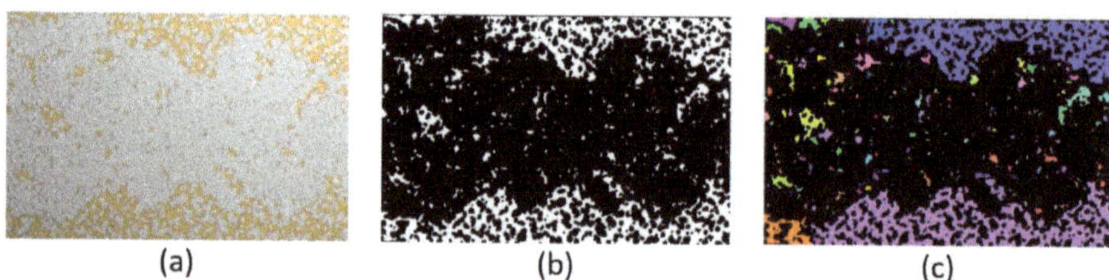

(a)   (b)   (c)

Figure 4. (**a**) Pre-processing image of a microchip (oil in gold, glass & brine in white), (**b**) a post-segmentation binary image (oil in white), and (**c**) oil clusters were labeled by pseudorandom colors for qualitative analysis. Quantitative analysis was applied to the binary image (**b**).

4. Results and Discussion

4.1. Fluid-Fluid Interactions

Interfacial tension of LSW-crude oil system was 19.2 (±0.02) and 15.3 (±0.03) mN/m for crude oil C and D, respectively. These are considered reference values for the specific oil type, to which the nanoparticles were compared. Crude oil D had fewer asphaltenes compared to crude oil C, and overall lower IFT values were obtained with crude oil D. This trend in IFT based on oil composition has also been reported in the literature [60].

The same tendency in IFT was seen for both oil types (Table 7), with 0.1 wt.% T-CNFs being similar to the reference value and 1.0 wt.% CNCs being lower. A low IFT is favorable, as it will increase the capillary number, which, in turn, will help to mobilize more oil. However, the decrease in IFT using CNCs was not of orders of magnitude, which is a necessary requirement for it to be considered a primary recovery mechanism [61]. Nevertheless, some studies have also shown that an ultralow IFT might not be necessary to improve the oil recovery [62].

4.2. Fluid-Solid Interactions

The contact angle is considered to be the most universal measurement of the wettability of a surface and it is an approach to measure reservoir wettability [26]. Wettability alteration has often been ascribed to the asphaltene content in crude oil, where crude oil with a high asphaltene content renders often more oil-wet surfaces [57]. However, this was not observed in the current study, as crude oil C yielded lower contact angle values compared to crude oil D (Table 7).

From Table 7 it is seen that the contact angle for LSW-crude oil-glass system, was 49.6° (±0.2°) and 52.5° (±0.8°) for crude oil C and D, respectively. These are considered reference values. Looking at crude oil C, all final values (12th hour) were similar to the reference value. For crude oil D, the addition of nanoparticles to the solution caused a small increase in the contact angle, with T-CNFs having the highest value. A higher value in the contact angle is equivalent to the system becoming less water-wet. However, in this case, the system was still in the water-wet regime.

Table 7. Interfacial tension (IFT) and contact angle values. Each experiment lasted for 12 h. The average value from the last hour of the experiment is reported in the table. Measurements were made at ambient conditions.

Fluid	Interfacial Tension (mN/m)		Contact Angle (°)	
	Crude Oil C	Crude Oil D	Crude Oil C	Crude Oil D
0.1 wt.% NaCl	19.2 ± 0.02	15.3 ± 0.03	49.6 ± 0.2	52.5 ± 0.8
1.0 wt.% CNCs	16.9 ± 0.02	13.8 ± 0.05	52.0 ± 0.1	56.1 ± 0.2
0.1 wt.% T-CNFs	19.1 ± 0.07	15.4 ± 0.03	51.1 ± 0.1	60.3 ± 0.1

4.3. Capillary Number

The capillary number for the core floods can be found in Table 8. A normal water flood has a capillary number that is in the order of 10^{-7} [63]. To mobilize the remaining oil, the capillary number should be in the range of 10^{-5} or higher [61]. For the low rate water and nano floods, the capillary number was 10^{-6}, while it was increased to 10^{-5} for the high rate floods. Thus, the increase in rate is expected to result in higher incremental oil production.

Looking at the effect of particle type and crude oil type, the capillary number was slightly higher for T-CNFs, compared to CNCs. In addition, crude oil D had slightly higher numbers than crude oil C, but they were all in the same order of magnitude.

Table 8. Overview of capillary numbers for core floods. * The reported capillary number is an average based on all the floods conducted with the respective fluid. For these floods, the capillary numbers were all in the same order of magnitude.

Fluid	Capillary Number for Core Floods			
	Crude Oil C		Crude Oil D	
	0.3 mL/min	3.0 mL/min	0.3 mL/min	3.0 mL/min
0.1 wt.% NaCl	1.23×10^{-6}	1.23×10^{-5}	1.6×10^{-6} *	1.6×10^{-5} *
1.0 wt.% CNCs	2.50×10^{-6}	2.50×10^{-5}	2.7×10^{-6} *	2.7×10^{-5} *
0.1 wt.% T-CNFs	-	-	6.0×10^{-6}	6.0×10^{-5}

4.4. Core Flood

4.4.1. Part 1

The objective of the secondary recovery floods was to see if CNC nanofluid produced more oil compared to low salinity water.

The trapping efficiency of a porous medium is significantly associated with the structure of the pore-space through factors such as pore body-throat aspect ratio and pore coordination number [64–67]. Therefore, the decision was taken to reutilize Core 1 for testing the 1 wt.% CNC nanofluid, via rinsing the core with methanol and toluene using a Soxhlet apparatus, and then drying the core at 60 °C. After drying, a new permeability and porosity measurement were obtained to compare with the original measurements (Table 9). The wettability restoration method applied to the core was considered successful, since it led to the same irreducible water saturation after primary drainage (Table 10). Although Core 1 permeability and porosity was reduced by 6.5% and 1.1%, respectively, it is assumed that the pore structure and wettability effect on oil recovery is minimized by using this approach.

Table 9. Pore volume of core 1 after cleaning, together with permeability and porosity data before and after water flood and core cleaning.

Core	New Pore Volume (mL)	Porosity (%)		Permeability (mD)		Reduction Perm. (%)
		Before	**After**	**Before**	**After**	
1	19.2	17.5	17.3	781	731	6.5

By looking at the pressure curves for secondary water- and CNC nano flood (Figure 5), the pressure was higher for the CNC nano flood stage, and it was slightly increasing for both the low rate and the high rate. However, no more oil was produced even though the pressure increased. Looking at the recovery, water flooding resulted in 52.0% total recovery, where 40.9% oil was produced during low rate and 11.1% oil was produced during the high rate (Table 10). For the second test using 1.0 wt.% CNC nanofluid (Test 1B), a higher recovery during the low rate (48.6%) and similar recovery to water flooding during the high rate (9.2%) were observed, overall resulting in a slightly higher total recovery of 57.8% oil (Table 10).

With this method, the nanofluid yielded 5.8% of original oil in place (OOIP)—more oil compared to low salinity water (Table 10). The increased oil recovery could either be a result of the viscosity difference between water and nanocellulose, or it could be a result of a physical interaction between the stiff and solid nanocellulose particles and the oil droplets. Another explanation for the increased oil recovery may be that pore throats could be blocked by the nanocellulose particles, leading the fluid in other directions.

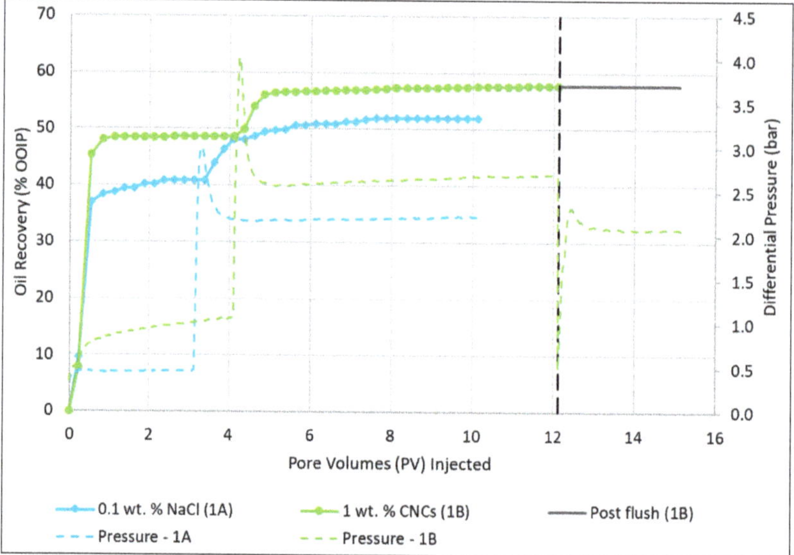

Figure 5. Oil recovery and pressure curves for secondary water flood (Test 1A) and secondary CNC nano flood (Test 1B). The black dotted vertical line indicates where post water flush was initiated for the nano flood. The spike in pressure curves indicates where the rate was switched from 0.3 mL/min to 3.0 mL/min, for the respective floods.

This approach was only performed on one core as it was a time-consuming process and uncertainties regarding the wettability of the core arose. Soxhlet extraction can sometimes change the core from a water-wet state to an oil-wet state. In addition, the solvent may not contact all of the core [68].

4.4.2. Part 2

In Part 2, CNC particles were used, and the parameters that were varied were wettability of the rock, temperature, and injection scheme. The results from this part are presented according to the injection mode.

Secondary Recovery

Further investigations of secondary recovery floodings were commenced by varying the wettability of the rock and temperature. In Figure 6, the recovery factors and permeability of the conducted experiments are illustrated. It is observed that CNC nanofluid produced more oil than LSW, when compared with the equal parallel experiment (Figure 6 and Table 10).

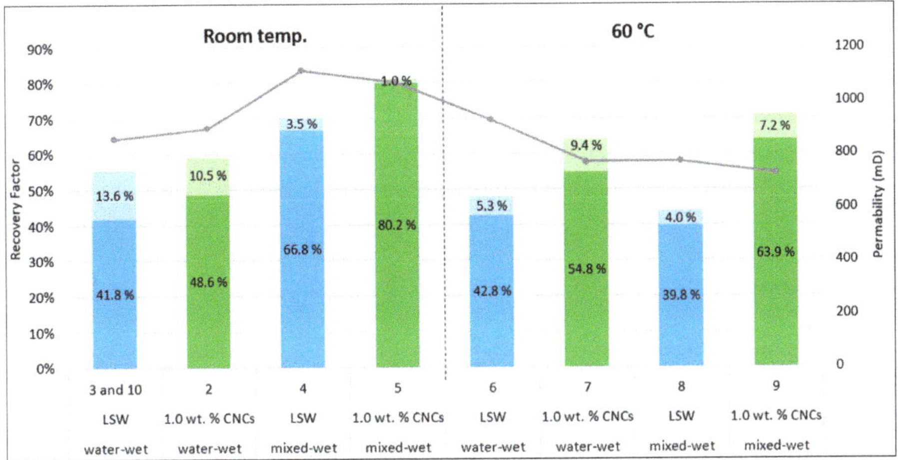

Figure 6. Recovery factors (% of original oil in place (OOIP)) and permeability of secondary recovery experiments. The vertical black dotted line acts as a marker between the experiments conducted at room temperature and 60 °C. Dark blue or dark green colors represent the recovery factor achieved from the low rate flooding (0.3 mL/min), while the lighter colors denote the recovery from the high rate flooding (3.0 mL/min). The numbers on the *x*-axis corresponds to the test number.

Four cores were aged in total, to make them more mixed-wet. Three cores were aged for five weeks (35 days); Test 4, 5, and 8, while Test 9 was aged for seven weeks (49 days). From an USBM wettability experiment Anderson [57] conducted on a Berea sandstone core, the wettability changed from water-wet (W = 0.8) to moderately oil wet (W = −0.3) over a 40-day period. However, after 35 days, the value was approximately −0.15, which indicates that the core was neutrally to moderately oil-wet. Thus, the chosen aging time in the current experimental study should be sufficiently long to alter the wettability to some degree. However, since it is uncertain if the wetting equilibrium has been reached, these cores will be considered intermediate or mixed-wet.

By looking at the effect of oil recovery and rock wettability, it seemed like the nanofluid was able to produce more oil when the core had been aged, and this was the case for both the room temperature and 60 °C flood. For LSW, on the other hand, no clear trend was observed, as the experiments at room temperature yielded a higher oil recovery for the aged core (Test 4), while, at 60 °C, more oil was produced with the water-wet core (Test 6). However, it should be noted that the permeability difference between the water-wet core and the mixed-wet core at 60 °C using LSW was approximately 150 mD. In addition, the total oil recovery was similar to each other—48.1% (Test 6) versus 43.8% (Test 8). A higher permeability for core 8 could, therefore, have resulted in a slightly higher recovery, which, in turn, could have given the same trend as was observed with the nanofluid.

All the room temperature floods resulted in a higher total recovery compared to their parallel experiment conducted at 60 °C, except for the nano flood in water-wet core (Test 2 compared to Test 7). However, in Test 2 crude oil C was used, as opposed to crude oil D, which was used for the other experiments. Test 4 and 5 were two of the experiments with the largest recovery during low rate flooding. However, they were also two of the cores with the highest permeability, which, in turn, could be a contributing factor to an increased oil recovery.

Tertiary Recovery

CNC nanofluid was introduced as a tertiary recovery technique, injected after water flooding, to see if incremental oil could be produced. With this injection mode, it would be more difficult for the nanoparticles to extract more oil, as most of the oil has already been recovered during water flooding. However, for all core floods, incremental oil was produced during the tertiary stage using CNC nanoparticles. The obtained results can be seen in Figure 7 and Table 10.

The four tests presented in Figure 7 are the same ones as already presented as secondary mode water floods in Figure 6. Thus, the focus of this part will be on the incremental oil production induced by the nanoparticles (green color in the figure).

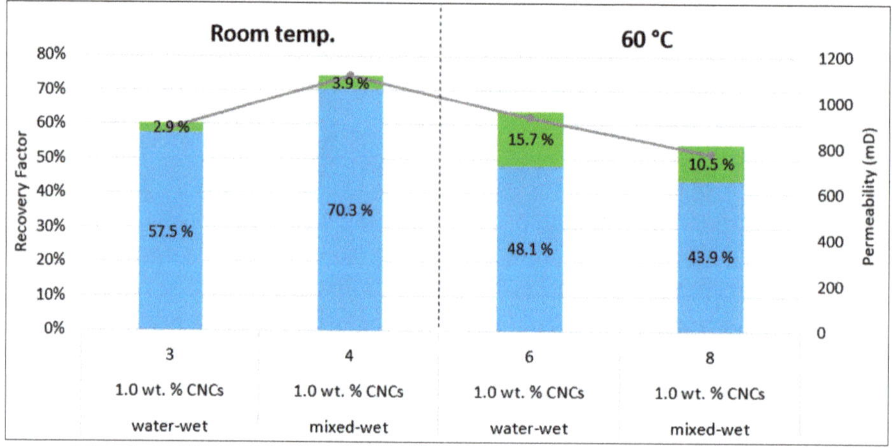

Figure 7. Recovery factors (% of OOIP) of tertiary recovery experiments. The vertical black dotted line acts as a marker between the experiments conducted at room temperature and 60 °C. Blue color represent the recovery factor achieved from low and high rate water flooding, while the green color denotes the recovery from the low and high rate nano flooding.

In terms of oil recovery and wettability, the temperature had an effect as to which wettability conditions gave the highest incremental oil recovery. For the room temperature floods, the highest incremental recovery was observed for the mixed-wet rock. The opposite was seen when the temperature was increased, with the water-wet rock yielding the highest incremental oil recovery. However, as previously discussed in the secondary mode results, there is a permeability difference of 150 mD between Test 6 and Test 8, which could be an attributing factor to the difference in recovery. In addition, by looking at all permeability data (Figure 7), there seems to be a correlation between permeability and recovery, as recovery is decreasing when permeability is decreasing. Thus, more experiments should be conducted with cores, using different permeabilities to be able to confirm that the temperature plays a role regarding the wettability preference and oil recovery. Consequently, from these experiments, no trend can be concluded at this point, as permeability might be the dominating factor in the oil recovery, but not the wettability.

Regarding the effect of temperature, the largest increase in recovery was observed for the floods conducted at 60 °C, where 15.7% (Test 6) and 10.5% (Test 8) additional oil was produced. However, it should be noted that the room temperature floods had a much higher secondary mode (LSW flood) recovery compared to the LSW floods at 60 °C. Thus, more oil was available for the nanoparticles when implementing the tertiary stage at 60 °C. Nevertheless, by looking at the viscosity difference between LSW and 1.0 wt.% CNCs (Table 3), the difference is larger at 60 °C, which could be a contributing factor as to why more oil was produced at the high temperature.

To summarize the secondary and tertiary floods (Part 2), CNC nanoparticles were able to extract 2–27% of OOIP more oil than LSW when injected as a secondary technique, and they appeared to perform better under mixed-wet conditions. The oil recovery was enhanced when CNC nanofluid was injected as a tertiary recovery technique, where more incremental oil was produced for the high-temperature floods. For the tertiary floods, there did not seem to be an overall trend regarding rock wettability and oil recovery, as permeability might have been the dominating factor to the resulting high oil recoveries.

4.4.3. Part 3

The effect of particle type was tested since CNCs have a different shape and size compared to T-CNFs (Figure 1). Thus, the two particles should result in a different behavior in the porous medium.

T-CNF and CNC Concentration

Two tertiary mode experiments were conducted, one using 0.1 wt.% T-CNFs (Test 10) and one using 1.0 wt.% CNCs (Test 3), both dispersed in 0.1 wt.% NaCl. The concentration of the two nanocellulose solutions differs from one another. A concentration of 0.1 wt.% T-CNFs was chosen, based on other flooding studies where T-CNFs have been employed [5].

CNCs were tested at different concentrations (0.5, 1.0, and 2.0 wt.%) in an injectivity study by Molnes et al. [7]. In their following work, they decided to move forward with the 0.5 wt.% CNCs, since it had the lowest differential pressure, but had still sufficiently effect on viscosity. The CNCs used in the current study are of the same quality as described by Molnes et al. [7]. Since 0.5 wt.% CNCs have already been documented in terms of EOR [6], it was of interest to test 1.0 wt.% CNCs, since it should give a higher viscosity effect, and might be able to recover more oil than the 0.5 wt.% CNCs. By looking at the tertiary mode experiments conducted in Part 2 in the current study, the CNC particles injected at 60 °C were attributable to an incremental oil recovery of 15.7% of OOIP and 10.5% of OOIP, respectively. No significant EOR effect was observed in the 60 °C experiment Molnes et al. [6] conducted using 0.5 wt.% CNCs.

A total of 0.1 wt.% CNCs were not considered a relevant concentration in the current experiments, as the effect of the particles was considered to be too low.

Oil Recovery Experiment

From Table 10, it is seen that the core flood using CNCs gave a total recovery of 60.4%, where the nanofluid contributed to a 2.9% incremental oil recovery. For the core flood where T-CNFs were employed, an additional oil recovery of 35.4% of OOIP oil was observed during the tertiary stage, yielding a total recovery of 88.7% of OOIP (Table 10). Based on the calculated recovery, T-CNFs seems more promising for EOR applications. However, the differential pressure kept increasing throughout the T-CNF nano flood (Figure 8), even though no oil was produced towards the end of the flood. A similar pressure trend has also been observed in a study performed by Kusanagi et al. [5], using another quality of T-CNFs.

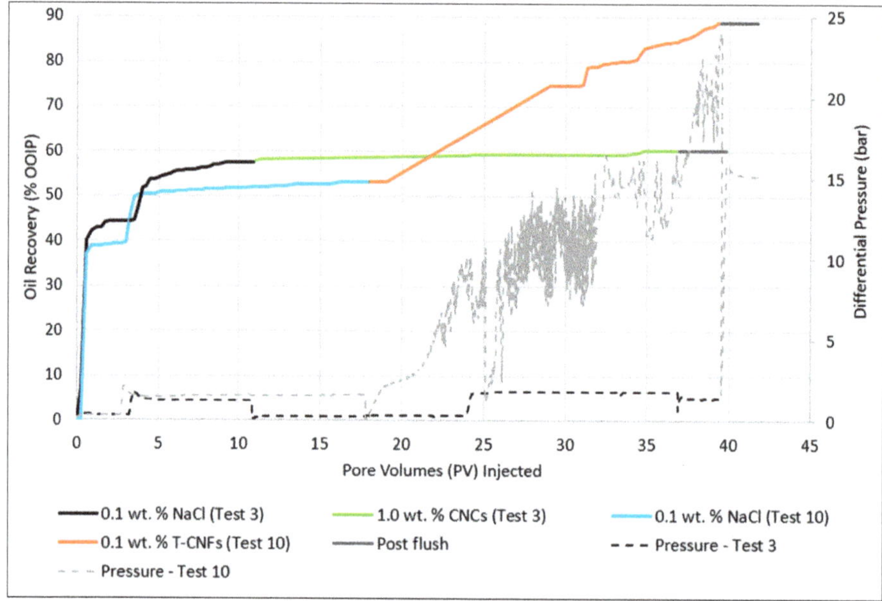

Figure 8. Recovery and corresponding pressure curves for 1.0 wt.% CNCs in 0.1 wt.% NaCl (Test 3) and 0.1 wt.% T-CNFs in 0.1 wt.% NaCl (Test 10).

For the flooding with CNCs, the differential pressure was relatively stable at the end of both the low and high rate nano flood, where it was around 0.3 bar and 1.8 bar, respectively. For T-CNFs, there was a large response in pressure when the nanofluid injection started. The pressure quickly increased, but towards the end of the low rate, it stabilized around 12 bar. The differential pressure of the high rate injection continuously increased and finally reached a maximum pressure of 24.1 bar, which was approximately 11 times higher than what was observed for the high rate nano flood using CNC particles. At 24.1 bar it was decided to stop the T-CNF nano flood experiment, as the differential pressure was approaching the value of the surrounding sleeve pressure. 21 PVs of T-CNFs had been injected at that point. The fluctuation in pressure (spikes) together with the constant increase could indicate that log-jammings were building up and breaking free inside the porous medium, or that the fluid had created a filter cake on the inlet side. Log-jams are a form of mechanical entrapment of particles, which occurs when particles accumulate at pore throats leading to a blocked pore and reduced permeability, which in turn results in an increased differential pressure [3]. T-CNFs have a big aspect ratio, so the accumulation can be quite severe compared to CNCs. A filter cake was observed on the inlet side of the core after the experiment was done, which further supports the theory about particles being retained in the porous medium.

Table 10. Summary of initial water saturations (S_{wi}) and recovery factors (as% OOIP) for the core floods for each of the flooding stages (low rate (Q_{low}), 0.3 mL/min and high rate (Q_{high}), 3.0 mL/min). * This test has been used for comparison with an experiment in Part 3.

| Part | Test No. | Fluids | | | | Conditions | | | Incremental Recovery Factor of OOIP (%) | | | | Total Recovery (%) |
| | | Secondary Agent | Tertiary Agent | Crude Oil Type | T | Aging Time | Swi | Secondary Agent | | Tertiary Agent | | |
					($^\circ$C)	(Weeks)	(Fraction)	Q_{low}	Q_{high}	Q_{low}	Q_{high}	
1	1A	0.1 wt.% NaCl	-	C		-	0.247	40.9	11.1	-	-	52.0
	1B	1.0 wt.% CNCs	-			-	0.241	48.6	9.2	-	-	57.8
2	2	1.0 wt.% CNCs	-		24	-	0.314	48.6	10.5	-	-	59.0
	3 *	0.1 wt.% NaCl	1.0 wt.% CNCs			-	0.306	44.3	13.2	1.7	1.2	60.4
	4	0.1 wt.% NaCl	1.0 wt.% CNCs			5	0.308	66.8	3.5	3.6	0.3	74.2
	5	1.0 wt.% CNCs	-			5	0.302	80.2	1.0	-	-	81.2
	6	0.1 wt.% NaCl	1.0 wt.% CNCs	D	60	-	0.407	42.8	5.3	9.4	6.3	63.8
	7	1.0 wt.% CNCs	-			-	0.383	54.8	9.4	-	-	64.2
	8	0.1 wt.% NaCl	1.0 wt.% CNCs			5	0.173	39.8	4.0	2.7	7.8	54.4
	9	1.0 wt.% CNCs	-			7	0.311	63.9	7.2	-	-	71.1
3	10	0.1 wt.% NaCl	0.1 wt.% T-CNFs		24	-	0.411	39.4	14.0	25.5	9.9	88.7
	3	0.1 wt.% NaCl	1.0 wt.% CNCs			-	0.306	44.3	13.2	1.7	1.2	60.4

4.5. Microfluidics

The oil recovery factor was 71.7%, 77.7%, and 93.2% for test M1 (0.1 wt.% NaCl), M2 (1.0 wt.% CNCs) and M3 (0.1 wt.% T-CNFs), respectively, as shown in Figure 9a.

The highest and lowest connectivity of the residual oil was observed in the CNC flood (X = −0.38) and T-CNF flood (X = −0.87), respectively (Figure 9b). In experiment M1, the LSW flood resulted in a residual oil connectivity of −0.63. This flood had a faster disconnecting rate, which was due to the poor sweep efficiency compared to the other two floods, and consequently, the oil recovery started earlier for LSW.

T-CNFs significantly reduced the size of the remaining oil clusters compared to CNCs and LSW, by breaking up large oil clusters and then mobilizing them. This was evident from the number of clusters and size of the remaining clusters shown in Figure 9c,d. The LSW flood resulted in almost the same number of clusters as the T-CNF flood, but the system had a high trapping efficiency of the largest and smallest oil clusters. The CNC flood showed a higher reduction in oil saturation by 6% with less fraction of large remaining oil clusters compared to the LSW flood. It significantly attributed to a better sweep efficiency. Figure 10 shows the change in oil distribution in the analyzed part of the microchip as a function of pore volume injected (PVI) for experiments M1, M2, and M3, where it can also be seen that T-CNFs performed better than the two other fluids in improving the oil recovery.

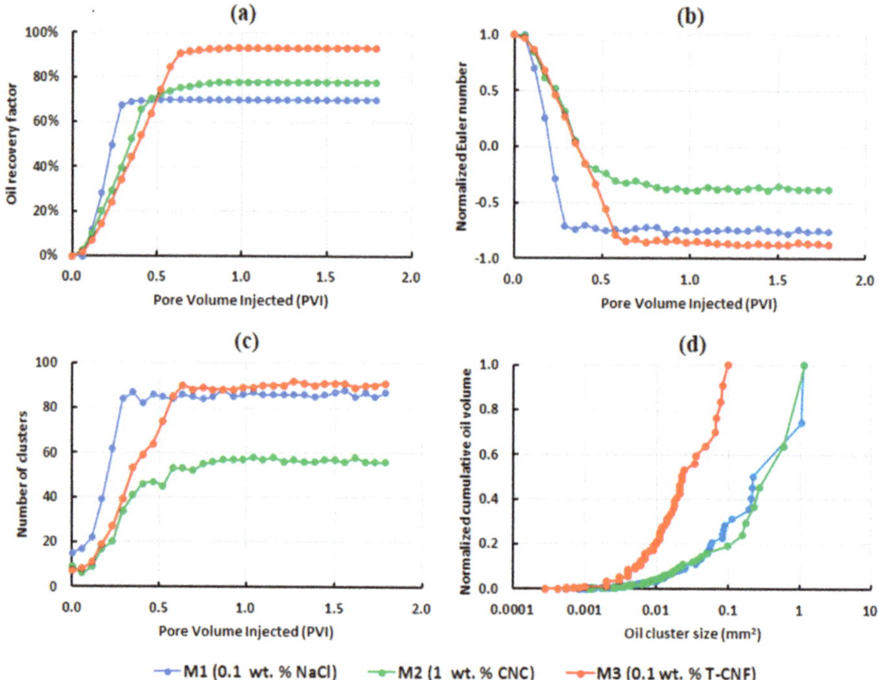

Figure 9. Results for the low rate injection microfluidic experiments: (**a**) oil recovery factor; (**b**) normalized Euler number; (**c**) number of clusters versus pore volume injected; and (**d**) normalized cumulative residual oil volume, as a function of oil cluster size.

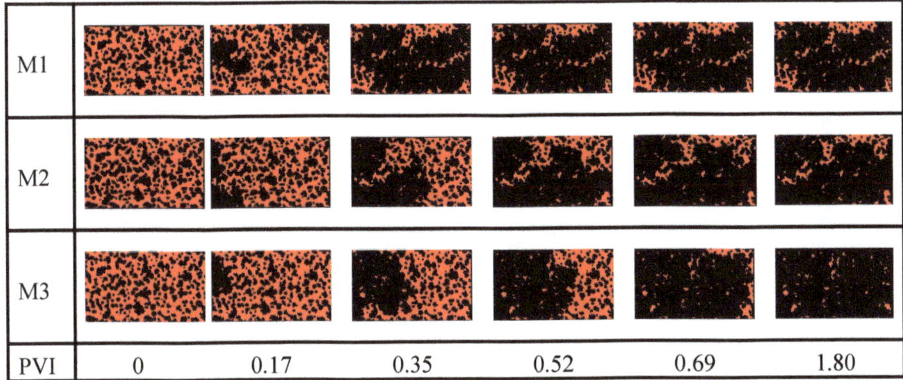

Figure 10. Change in the oil (in red) configuration in the analyzed part of the microchip as a function of pore volume injected (PVI) for experiment M1, M2, and M3.

The effect of flow rate was tested by injecting CNCs (M4) and T-CNFs (M5). The applied flow rate was 10 times higher than the rate used in the corresponding parallel experiment, M2 (CNCs) and M3 (T-CNFs), respectively. A higher rate resulted in lower oil recovery factors and higher remaining oil connectivity as shown in Figure 11a,b. In addition, it was found that increasing the injection rate increased the residual number of clusters produced by T-CNFs, while the CNC flood led to a lower number of clusters, as illustrated in Figure 11c.

Figure 12 illustrates the post-flooding images corresponding to experiments M2–5, and it is clear that increasing the flow rate resulted in trapping of larger oil clusters, particularly for CNCs, where a very large cluster remained in the analyzed part of the microchip.

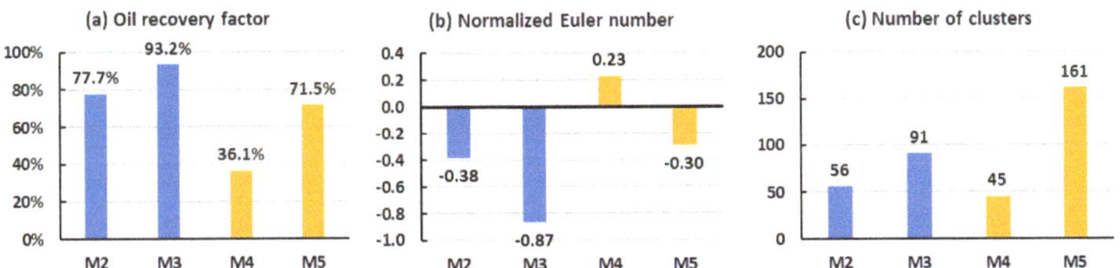

Figure 11. (**a**) Oil recovery, (**b**) normalized Euler number, and (**c**) number of clusters in the post-flooding images of low rate (in blue) and high rate (in yellow) microfluidic experiments. CNCs were used in M2 and M4, and T-CNFs were used in M3 and M5.

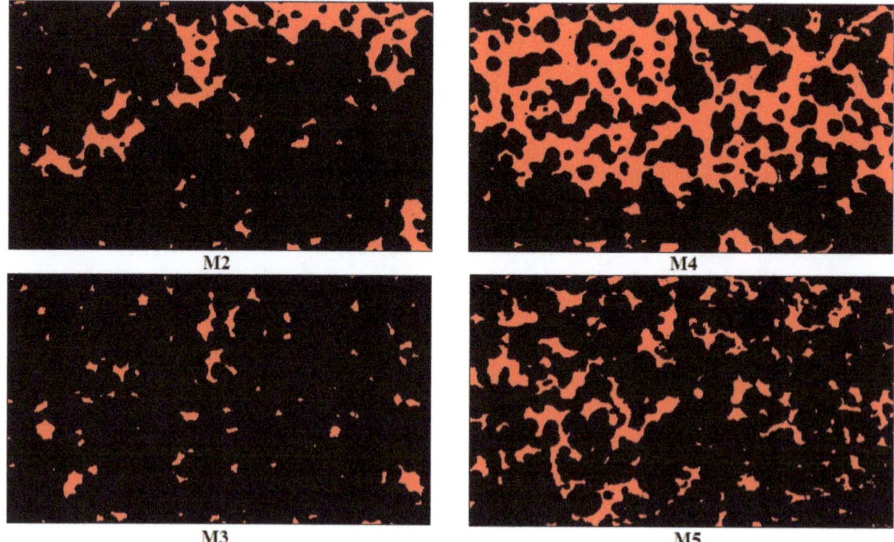

Figure 12. Illustration of the reimaging oil post flooding in CNCs experiments (M2, M4) and T-CNFs experiments (M3, M5). M4 and M5 were performed with 10 times higher injection rate, i.e., at 1.8 μL/min vs. 0.18 μL/min flow rate.

5. Conclusions

In this study, ten core floods and five microfluidic experiments were conducted to investigate the potential nanocelluloses have as a chemical EOR additive in low salinity water. Two types of nanocellulose particles were tested: cellulose nanocrystals and TEMPO-oxidized cellulose nanofibrils. In the study, the effect of injection scheme, temperature, wettability, and particle type were looked upon. In addition, interfacial tension and contact angle measurements were conducted to further support the findings from the floods. Based on the study the following conclusions can be drawn:

a. The interfacial tension and contact angle values were dependent upon the crude oil type and nanoparticle type that was used. Overall, the IFT was not altered by the addition of T-CNF nanoparticles, but a small decrease in IFT was observed when CNCs were employed. For the contact angle, a slight increase in value was observed when CNCs or T-CNFs were added to the LSW. However, the change was marginal, thus, wettability alteration is not a primary EOR mechanism.

b. For the secondary mode experiment where the core was re-used, the wettability restoration method was considered successful since it led to the same irreducible water saturation after primary drainage. From the oil recovery experiment, nanofluid yielded 5.8% of OOIP more oil, compared to low salinity water.

c. CNC nanoparticles were able to extract 2–27% of OOIP more oil than LSW when injected as a secondary technique. Furthermore, the particles appeared to perform better under mixed-wet conditions.

d. The oil recovery was enhanced when CNC nanofluid was injected as a tertiary recovery technique, where more incremental oil was produced for the high-temperature floods. For the tertiary floods, there did not seem to be an overall trend regarding rock wettability and oil recovery.

e. Looking at the effect of particle type, T-CNFs were much more effective than CNCs to recover trapped oil. This was evident from both core flooding and microfluidic experiments. However, the pressure was constantly increasing during the high rate T-CNF core flood. Even though more oil was recovered during the T-CNF flood, the high pressure indicates poor injectivity. Furthermore, filtering of particles was observed on the inlet side of the core plug after the experiment. Future experiments should, therefore, test T-CNFs at a lower concentration, to see if a similar high incremental oil recovery can be achieved with a lower and more stable pressure profile.

f. The microfluidic experiments supported the findings from the core floods, with nanofluid leading to a better sweep efficiency compared to low salinity flooding. T-CNFs improved the oil recovery the most, by breaking up large oil clusters and mobilizing them. Looking at the effect of the flow rate, it was evident that a higher flow rate resulted in lower oil recovery factors and higher remaining oil connectivity.

Author Contributions: Conceptualization, R.C.A.; methodology, data curation, formal analysis, investigation and validation, R.C.A. and S.A.; funding acquisition, project administration, resources and supervision, E.B.H., K.S. and O.T.; visualization and writing—original draft preparation, R.C.A. and S.A.; writing—review and editing, R.C.A., S.A., E.B.H., K.S. and O.T. All authors have read and agreed to the published version of the manuscript.

Funding: This research was funded by the Research Council of Norway through grant 244615/E30 in the Petromaks2 program, and trough the Centres of Excellence funding scheme, project number 262644.

Acknowledgments: The authors would like to thank the Research Council of Norway for their financial support through the GreenEOR project (grant 244615/E30) in the Petromaks2 program, and through the Centres of Excellence funding scheme, project number 262644. The authors would also like to thank master students Yuntian Teng and Hang Bian for collaboration on the core flood experiments. Thank you to NTNU laboratory engineer Roger Overå for assistance, and RISE PFI engineers Ingebjørg Leirset and Mirjana Filipovic for their work with production of the TEMPO-oxidized CNFs and Per Olav Johnsen for acquiring AFM images.

Conflicts of Interest: The authors declare no conflict of interest.

References

1. Muggeridge, A.; Cockin, A.; Webb, K.; Frampton, H.; Collins, I.; Moulds, T.; Salino, P. Recovery rates, enhanced oil recovery and technological limits. *Philos. Trans. R. Soc. A* **2014**, *372*, 20120320. [CrossRef]

2. Available online: https://www.regjeringen.no/globalassets/upload/oed/pdf_filer_2/faktaheftet/fakta2014og/facts_2014_nett_.pdf (accessed on 7 April 2014).

3. Aadland, R.C.; Jakobsen, T.D.; Heggset, E.B.; Long-Sanouiller, H.; Simon, S.; Paso, K.G.; Syverud, K.; Torsæter, O. High-temperature core flood investigation of nanocellulose as a green additive for enhanced oil recovery. *Nanomaterials* **2019**, *9*, 665. [CrossRef] [PubMed]

4. Jakobsen, T.D.; Simon, S.b.; Heggset, E.B.; Syverud, K.; Paso, K.J.I.; Research, E.C. Interactions between surfactants and cellulose nanofibrils for enhanced oil recovery. *Ind. Eng. Chem. Res.* **2018**, *57*, 15749–15758. [CrossRef]

5. Kusanagi, K.; Murata, S.; Goi, Y.; Sabi, M.; Zinno, K.; Kato, Y.; Togashi, N.; Matsuoka, T.; Liang, Y. Application of cellulose nanofiber as environment-friendly polymer for oil development. In Proceedings of the SPE/IATMI Asia Pacific Oil & Gas Conference and Exhibition, Nusa Dua, Bali, Indonesia, 20–22 October 2015.

6. Molnes, S.N.; Mamonov, A.; Paso, K.G.; Strand, S.; Syverud, K. Investigation of a new application for cellulose nanocrystals: A study of the enhanced oil recovery potential by use of a green additive. *Cellulose* **2018**, *25*, 2289–2301. [CrossRef]

7. Molnes, S.N.; Torrijos, I.P.; Strand, S.; Paso, K.G.; Syverud, K. Sandstone injectivity and salt stability of cellulose nanocrystals (CNC) dispersions—Premises for use of CNC in enhanced oil recovery. *Ind. Crop. Prod.* **2016**, *93*, 152–160. [CrossRef]

8. French, A.; Bertoniere, N.; Brown, R.; Chanzy, H.; Gray, D.; Hattori, K.; Kirk-Othmer, G. *Encyclopedia of Chemical Technology*; Seidel, A., Ed.; John Wiley Sons, Inc.: Hoboken, NJ, USA, 2004; Volume 5, pp. 360–394.

9. Krässig, H.A. *Cellulose: Structure, Accessibility and Reactivity*; Gordon and Breach Science Publ.: Amsterdam, The Netherlands, 1993.

10. Klemm, D.; Kramer, F.; Moritz, S.; Lindström, T.; Ankerfors, M.; Gray, D.; Dorris, A. Nanocelluloses: A new family of nature-based materials. *Angew. Chem. Int. Ed.* **2011**, *50*, 5438–5466. [CrossRef] [PubMed]

11. Habibi, Y.; Lucia, L.A.; Rojas, O.J. Cellulose nanocrystals: Chemistry, self-assembly, and applications. *Chem. Rev.* **2010**, *110*, 3479–3500. [CrossRef] [PubMed]

12. Available online: https://www.ospar.org/work-areas/oic/chemicals (accessed on 29 May 2020).

13. Sheng, J.J.; Leonhardt, B.; Azri, N. Status of Polymer-Flooding Technology. *J. Can. Pet. Technol.* **2015**, *54*, 116–126. [CrossRef]

14. Heggset, E.B.; Chinga-Carrasco, G.; Syverud, K. Temperature stability of nanocellulose dispersions. *Carbohydr. Polym.* **2017**, *157*, 114–121. [CrossRef] [PubMed]

15. Wei, B.; Li, Q.; Jin, F.; Li, H.; Wang, C. The Potential of a Novel Nanofluid in Enhancing Oil Recovery. *Energy Fuels* **2016**, *30*, 2882–2891. [CrossRef]

16. Wei, B.; Li, Q.; Ning, J.; Wang, Y.; Sun, L.; Pu, W. Macro-and micro-scale observations of a surface-functionalized nanocellulose based aqueous nanofluids in chemical enhanced oil recovery (C-EOR). *Fuel* **2019**, *236*, 1321–1333. [CrossRef]

17. Bila, A.; Stensen, J.Å.; Torsæter, O. Experimental Investigation of Polymer-Coated Silica Nanoparticles for Enhanced Oil Recovery. *Nanomaterials* **2019**, *9*, 822. [CrossRef] [PubMed]

18. Nasralla, R.A.; Nasr-El-Din, H.A. Double-layer expansion: Is it a primary mechanism of improved oil recovery by low-salinity waterflooding? *SPE Reserv. Eval. Eng.* **2014**, *17*, 49–59. [CrossRef]

19. Rivet, S.; Lake, L.W.; Pope, G.A. A coreflood investigation of low-salinity enhanced oil recovery. In Proceedings of the SPE Annual Technical Conference and Exhibition, Florence, Italy, 20–22 September 2010.

20. Shaker Shiran, B.; Skauge, A. Enhanced oil recovery (EOR) by combined low salinity water/polymer flooding. *Energy Fuels* **2013**, *27*, 1223–1235. [CrossRef]

21. Siyambalagoda Gamage, P.H.; Thyne, G.D. Comparison of Oil Recovery by Low Salinity Waterflooding in Secondary and Tertiary Recovery Modes. In Proceedings of the SPE Annual Technical Conference and Exhibition, Denver, CO, USA, 1 January 2011.

22. Zhang, Y.; Morrow, N.R. Comparison of secondary and tertiary recovery with change in injection brine composition for crude-oil/sandstone combinations. In Proceedings of the SPE/DOE Symposium on Improved Oil Recovery, Tulsa, OK, USA, 22–26 April 2006.

23. Anderson, W.G. Wettability literature survey-part 6: The effects of wettability on waterflooding. *Pet. Technol.* **1987**, *39*, 1605–1622. [CrossRef]

24. Dandekar, A.Y. *Petroleum Reservoir Rock and Fluid Properties*; CRC press: New York, NY, USA, 2013.

25. Jadhunandan, P.; Morrow, N.R. Effect of wettability on waterflood recovery for crude-oil/brine/rock systems. *SPE Reserv. Eng.* **1995**, *10*, 40–46. [CrossRef]

26. Morrow, N.R. Wettability and its effect on oil recovery. *Pet. Technol.* **1990**, *42*, 1476–1484. [CrossRef]

27. Rücker, M.; Bartels, W.-B.; Singh, K.; Brussee, N.; Coorn, A.; van der Linde, H.A.; Bonnin, A.; Ott, H.; Hassanizadeh, S.M.; Blunt, M.J.; et al. The effect of mixed wettability on pore-scale flow regimes based on a flooding experiment in Ketton Limestone. *Geophys. Res. Lett.* **2019**, *46*, 3225–3234.

28. Treiber, L.; Owens, W. A laboratory evaluation of the wettability of fifty oil-producing reservoirs. *Soc. Pet. Eng. J.* **1972**, *12*, 531–540. [CrossRef]

29. Ashraf, A.; Hadia, N.; Torsaeter, O.; Tweheyo, M.T. Laboratory investigation of low salinity waterflooding as secondary recovery process: Effect of wettability. In Proceedings of the SPE Oil and Gas India Conference and Exhibition, Mumbai, India, 1 January 2010.

30. Gamage, P.; Thyne, G. Systematic investigation of the effect of temperature during aging and low salinity flooding of berea sandstone and minn. In Proceedings of the 16th European Symposium on Improved Oil Recovery, Cambridge, UK, 12–14 April 2011.

31. Thomas, S. Enhanced Oil Recovery-An Overview. *Oil Gas. Sci. Technol.-Revue de l'IFP* **2008**, *63*, 9–19. [CrossRef]

32. Foster, W.R. A Low-Tension Waterflooding Process. *Pet. Technol.* **1973**, *25*, 205–210. [CrossRef]

33. Zolotukhin, A.B.; Ursin, J.-R. *Introduction to Petroleum Reservoir Engineering*; Høyskoleforlaget (Norwegian Academic Press): Kristiansand, Norway, 2000.

34. Khishvand, M.; Akbarabadi, M.; Piri, M. Micro-scale experimental investigation of the effect of flow rate on trapping in sandstone and carbonate rock samples. *Adv. Water Resour.* **2016**, *94*, 379–399. [CrossRef]

35. Zhang, H.; Ramakrishnan, T.S.; Nikolov, A.; Wasan, D. Enhanced oil recovery driven by nanofilm structural disjoining pressure: Flooding experiments and microvisualization. *Energy Fuels* **2016**, *30*, 2771–2779. [CrossRef]

36. Hendraningrat, L.; Li, S.; Torsater, O. Effect of some parameters influencing enhanced oil recovery process using silica nanoparticles: An experimental investigation. In Proceedings of the SPE Reservoir Characterization and Simulation Conference and Exhibition, Abu Dhabi, UAE, 16 September 2013.

37. Cheraghian, G. An experimental study of surfactant polymer for enhanced heavy oil recovery using a glass micromodel by adding nanoclay. *Pet. Sci. Technol.* **2015**, *33*, 1410–1417. [CrossRef]

38. Herbas, J.; Wegner, J.; Hincapie, R.; Födisch, H.; Ganzer, L.; Del Castillo, J.; Mugizi, H.M. Comprehensive micromodel study to evaluate polymer EOR in unconsolidated sand reservoirs. In Proceedings of the SPE Middle East Oil & Gas Show and Conference, Manama, Bahrain, 8–11 March 2015.

39. Li, S.; Torsæter, O. An experimental investigation of EOR mechanisms for nanoparticles fluid in glass micromodel. In Proceedings of the International Symposium of the Society of Core Analysts, Avignon, France, 8–11 September 2014.

40. Bila, A.; Stensen, J.Å.; Torsæter, O. Experimental evaluation of oil recovery mechanisms using a variety of surface-modified silica nanoparticles in the injection water. In Proceedings of the SPE Norway One Day Seminar, Bergen, Norway, 13 May 2019.

41. Hosseini, S.J.; Foroozesh, J. Experimental study of polymer injection enhanced oil recovery in homogeneous and heterogeneous porous media using glass-type micromodels. *Pet. Explor. Prod. Technol.* **2019**, *9*, 627–637. [CrossRef]

42. Nilsson, M.A.; Kulkarni, R.; Gerberich, L.; Hammond, R.; Singh, R.; Baumhoff, E.; Rothstein, J.P. Effect of fluid rheology on enhanced oil recovery in a microfluidic sandstone device. *J. Non-Newton. Fluid Mech.* **2013**, *202*, 112–119. [CrossRef]

43. Joekar Niasar, V.; Hassanizadeh, S.; Pyrak-Nolte, L.; Berentsen, C. Simulating drainage and imbibition experiments in a high-porosity micromodel using an unstructured pore network model. *Water Resour. Res.* **2009**, *45*. [CrossRef]

44. Iglauer, S.; Fernø, M.; Shearing, P.; Blunt, M. Comparison of residual oil cluster size distribution, morphology and saturation in oil-wet and water-wet sandstone. *Colloid Interface Sci.* **2012**, *375*, 187–192. [CrossRef]

45. Zhao, J.; Wen, D. Pore-scale simulation of wettability and interfacial tension effects on flooding process for enhanced oil recovery. *RSC Adv.* **2017**, *7*, 41391–41398. [CrossRef]

46. Khanamiri, H.H.; Torsæter, O. Fluid topology in pore scale two-phase flow imaged by synchrotron X-ray microtomography. *Water Resour. Res.* **2018**, *54*, 1905–1917. [CrossRef]

47. Sivanesapillai, R.; Steeb, H. Fluid interfaces during viscous-dominated primary drainage in 2D micromodels using pore-scale SPH simulations. *Geofluids* **2018**, *2018*. [CrossRef]

48. Toriwaki, J.; Yoshida, H. *Fundamentals of Three-Dimensional Digital Image Processing*; Springer Science & Business Media: London, UK, 2009.

49. Dong, X.M.; Revol, J.-F.; Gray, D.G. Effect of microcrystallite preparation conditions on the formation of colloid crystals of cellulose. *Cellulose* **1998**, *5*, 19–32. [CrossRef]

50. Isogai, A.; Saito, T.; Fukuzumi, H. TEMPO-oxidized cellulose nanofibers. *Nanoscale* **2011**, *3*, 71–85. [CrossRef] [PubMed]

51. Araki, J.; Wada, M.; Kuga, S. Steric stabilization of a cellulose microcrystal suspension by poly (ethylene glycol) grafting. *Langmuir* **2001**, *17*, 21–27. [CrossRef]

52. Saito, T.; Isogai, A. TEMPO-mediated oxidation of native cellulose. The effect of oxidation conditions on chemical and crystal structures of the water-insoluble fractions. *Biomacromolecules* **2004**, *5*, 1983–1989. [CrossRef]

53. Aadland, R.C.; Dziuba, C.; Heggset, E.; Syverud, K.; Torsæter, O.; Holt, T.; Gates, I.; Bryant, S. Identification of Nanocellulose Retention Characteristics in Porous Media. *Nanomaterials* **2018**, *8*, 547. [CrossRef]

54. Aadland, R.C. Experimental study of flow of nanocellulose in porous media for enhanced oil recovery application. Ph.D Thesis, Norwegian University of Science and Technology, Trondheim, Norway, 2019.

55. Reid, M.S.; Villalobos, M.; Cranston, E.D. Benchmarking cellulose nanocrystals: From the laboratory to industrial production. *Langmuir* **2017**, *33*, 1583–1598. [CrossRef]

56. Sacui, I.A.; Nieuwendaal, R.C.; Burnett, D.J.; Stranick, S.J.; Jorfi, M.; Weder, C.; Foster, E.J.; Olsson, R.T.; Gilman, J.W. Comparison of the properties of cellulose nanocrystals and cellulose nanofibrils isolated from bacteria, tunicate, and wood processed using acid, enzymatic, mechanical, and oxidative methods. *ACS Appl. Mater. Interfaces* **2014**, *6*, 6127–6138. [CrossRef]

57. Anderson, W.G. Wettability literature survey-part 1: Rock/oil/brine interactions and the effects of core handling on wettability. *Pet. Technol.* **1986**, *38*, 1125–1144. [CrossRef]

58. Schindelin, J.; Arganda-Carreras, I.; Frise, E.; Kaynig, V.; Longair, M.; Pietzsch, T.; Preibisch, S.; Rueden, C.; Saalfeld, S.; Schmid, B. Fiji: An open-source platform for biological-image analysis. *Nat. Methods* **2012**, *9*, 676. [CrossRef] [PubMed]

59. Doube, M.; Kłosowski, M.M.; Arganda-Carreras, I.; Cordelières, F.P.; Dougherty, R.P.; Jackson, J.S.; Schmid, B.; Hutchinson, J.R.; Shefelbine, S.J. BoneJ: Free and extensible bone image analysis in ImageJ. *Bone* **2010**, *47*, 1076–1079. [CrossRef]

60. Tang, G.-Q.; Morrow, N.R. Salinity, temperature, oil composition, and oil recovery by waterflooding. *SPE Reserv. Eng.* **1997**, *12*, 269–276. [CrossRef]

61. Lake, L.W. *Enhanced oil recovery*; Prentice-Hall, Inc.: Upper Saddle River, NJ, USA, 1989.

62. Raffa, P.; Broekhuis, A.A.; Picchioni, F. Polymeric surfactants for enhanced oil recovery: A review. *Pet. Sci. Eng.* **2016**, *145*, 723–733. [CrossRef]

63. Skjæveland, S.M.; Kleppe, J. *SPOR Monograph-Recent Advances in Improved Oil Recovery Methods for North Sea Sandstone Reservoirs*; Norwegian Petroleum Directorate: Stavanger, Norway, 1992.

64. Blunt, M.J. *Multiphase Flow in Permeable Media: A Pore-Scale Perspective*; Cambridge University Press: Cambridge, UK, 2017.

65. Herring, A.; Robins, V.; Sheppard, A. Topological Persistence for Relating Microstructure and Capillary Fluid Trapping in Sandstones. *Water Resour. Res.* **2018**, *55*, 555–573. [CrossRef]

66. Mahmud, W.M.; Nguyen, V.H. Effects of snap-off in imbibition in porous media with different spatial correlations. *Transport. Porous Media* **2006**, *64*, 279–300. [CrossRef]

67. Tanino, Y.; Blunt, M.J. Capillary trapping in sandstones and carbonates: Dependence on pore structure. *Water Resour. Res.* **2012**, *48*. [CrossRef]

68. Gant, P.L.; Anderson, W.G. Core cleaning for restoration of native wettability. *SPE Form. Eval.* **1988**, *3*, 131–138. [CrossRef]

nanomaterials

Article

Insights into the Effects of Pore Size Distribution on the Flowing Behavior of Carbonate Rocks: Linking a Nano-Based Enhanced Oil Recovery Method to Rock Typing

Amin Rezaei [1], Hadi Abdollahi [2], Zeinab Derikvand [1], Abdolhossein Hemmati-Sarapardeh [3,4], Amir Mosavi [5,6,7,*] and Narjes Nabipour [8]

[1]   Abdal Industrial Projects Management Co. (MAPSA), Tehran 1456914477, Iran;
      arezaei@parspetro.com (A.R.); Z.derikvand@mapsaeng.com (Z.D.)
[2]   Department of Petroleum Engineering, Science and Research Branch, Azad University,
      Tehran 1477893855, Iran; habdollahi@srbiau.ac.ir
[3]   Department of Petroleum Engineering, Shahid Bahonar University of Kerman, Kerman 7616913439, Iran;
      hemmati@uk.ac.ir
[4]   College of Construction Engineering, Jilin University, Changchun 130600, China
[5]   Faculty of Civil Engineering, Technische Universität Dresden, 01069 Dresden, Germany
[6]   Kalman Kando Faculty of Electrical Engineering, Obuda University, 1034 Budapest, Hungary
[7]   Department of Mathematics, J. Selye University, 94501 Komarno, Slovakia
[8]   Institute of Research and Development, Duy Tan University, Da Nang 550000, Vietnam;
      narjesnabipour@duytan.edu.vn
*   Correspondence: amir.mosavi@mailbox.tu-dresden.de

Received: 18 March 2020; Accepted: 8 May 2020; Published: 18 May 2020

Abstract: As a fixed reservoir rock property, pore throat size distribution (PSD) is known to affect the distribution of reservoir fluid saturation strongly. This study aims to investigate the relations between the PSD and the oil–water relative permeabilities of reservoir rock with a focus on the efficiency of surfactant–nanofluid flooding as an enhanced oil recovery (EOR) technique. For this purpose, mercury injection capillary pressure (MICP) tests were conducted on two core plugs with similar rock types (in respect to their flow zone index (FZI) values), which were selected among more than 20 core plugs, to examine the effectiveness of a surfactant–nanoparticle EOR method for reducing the amount of oil left behind after secondary core flooding experiments. Thus, interfacial tension (IFT) and contact angle measurements were carried out to determine the optimum concentrations of an anionic surfactant and silica nanoparticles (NPs) for core flooding experiments. Results of relative permeability tests showed that the PSDs could significantly affect the endpoints of the relative permeability curves, and a large amount of unswept oil could be recovered by flooding a mixture of the alpha olefin sulfonate (AOS) surfactant + silica NPs as an EOR solution. Results of core flooding tests indicated that the injection of AOS + NPs solution in tertiary mode could increase the post-water flooding oil recovery by up to 2.5% and 8.6% for the carbonate core plugs with homogeneous and heterogeneous PSDs, respectively.

Keywords: nanomaterials; pore throat size distribution; mercury injection capillary pressure; interfacial tension; contact angle; enhanced oil recovery; surfactant; nanoparticle

---

1. Introduction

Residual oil remains in the reservoirs after conventional water flooding is often regarded as the target for enhanced oil recovery (EOR) processes. A comprehensive understanding and evaluation of the displacement efficiency represent a fundamental requirement for oil production forecast and field development planning [1–3]. At a laboratory scale, similar rock types have different residual oil saturations after water flooding, and a particular relation between the macroscopic characteristics and the flow behavior of the porous medium is yet to be developed [4,5]. Wardlaw et al. presented a low-cost method for estimating the amount of residual oil in a carbonate reservoir based on the observational evaluation of rock sections and pore casts [6]. They also stated that the oil recovery increases with the pore-to-throat diameter ratio, although this hypothesis is yet to be tested experimentally. Magara placed emphasis on the significant contribution of pore size to the oil production efficiency [7]. Chatzis et al. studied the effect of pore size on the residual oil saturation and ended up finding no significant association between the residual oil saturation and the rock pore size, although they observed an increase in the amount of the trapped oil at higher aspect ratios [8]. Elgaghah et al. considered the influence of pore size distribution (PSD) on the displacement efficiency in oil-bearing reservoirs [9]. They employed the image processing technique to find the pore size distribution of the rock samples in order to design the optimum microbial EOR process regarding the PSD of the porous media.

The amount of oil remained in the reservoir after the conventional water flooding was found to be dependent on the porosity and permeability of the porous medium [10], both of which are known to be influenced by the PSD of the reservoir rock [11]. Hence, it is essential to study the effect of PSD on the flow behavior of the reservoir rock to understand the governing mechanisms of the fluid flow through the porous medium. Al-Shalabi and Ghosh studied the effect of permeability contrast on the efficiency of water flooding in different porous media consisting of a glass micromodel (representing a high-permeability medium) and a set of sandstone core samples (representing a low- to medium-permeability medium) [1]. They employed three techniques to measure the water displacement efficiency, namely linear core flooding, micro model flooding, and imbibition by ultracentrifugation. Interestingly, their results showed a higher ultimate oil recovery in the lower-permeability medium. Gharibshahi et al. investigated the effects of pore shape, pore connectivity, pore heterogeneity, and tortuosity on the enhanced oil recovery (EOR) and the breakthrough time of micro models of nanofluid flow using computational fluid dynamics (CFD) [12]. Based on their results, they argued that a random distribution model tended to resemble the fluid flow in a hydrocarbon reservoir through the trapping effect on the flowing behavior that could not be investigated by such a model. However, to the best of our knowledge, there are only limited research works considering the effect of nanoscale confinement on the flow behavior of the fluid phase for water flooding in carbonate rocks. So far, there have been many studies elaborating on the effects of pore geometry on the permeability in the geological literature [13,14], but most of the relevant works have been focused on the single-phase fluid flow [15]. To better clarify the effects of PSD on the flow behavior of a reservoir rock, it is important to collect samples of similar rock types. Conventional reservoir rock typing has been defined as the classification of the reservoir rock based on its petrophysical properties acquired through wireline logs, porosity-permeability correlations, mercury injection curves, and geological features [5,16–20]. As a basis for reservoir rock classification, the pore geometry can be assessed based on the capillary pressure, which, in turn, can be better measured by the mercury injection analysis rather than other methods [21]. Most of the aforementioned research works have pointed out the relationship between the PSD and the residual oil saturation based on the analyses performed on a synthetic structure (i.e., micro model or sand-packs), making their results unreliable for using in actual reservoir studies. Accordingly, the present research is an attempt to fill in the research gap of investigating the effects of the PSD on the flow behavior of a porous medium.

Surfactant solution injection has been recommended as an efficient method to improve the oil displacement efficiency, provided the injection amount is sufficient, particularly for depleted reservoirs [22]. The surfactants have been widely studied as additives to decrease the oil–water interfacial tension (IFT) and residual oil saturation [23–30]. They also act as a high-grade agent to alter the rock surface wettability [31,32]. The ability of surfactant solutions to alter the wettability of solid surfaces depends on the properties of both the surfactant and the neighboring reservoir rock [33].

The primary effect of the surfactants that contributes to enhanced oil recovery is the reduction of the oil-brine IFT, even at low concentrations [26,27]. Adsorption of surfactants on the rock surface, on the other hand, reduces the efficiency of the EOR process [30,34]. So, in the present study, attempts were made to minimize the surfactant loss upon the adsorption on the rock surface.

Additionally, in recent years, several scholars have studied the feasibility of applying nanotechnology to different aspects of the petroleum industry [35–41]. Reservoir engineers have implemented nanoparticles (NPs) for different purposes in the petroleum industry, including the EOR. Surfactant-silica NPs solutions have drawn attention for their application in the EOR process, where they can help decrease the surfactant loss, reduce the NP dosage, increase the particle stability, and enhance the efficiency [42,43]. Roustaei obtained an additional 10.0% original oil in place (% OOIP) by imbibition of CTAB solution in 6.8 wt.% saline water at ambient conditions [44]. By integrating TX-100 (nonionic surfactant) with silica NPs, Zhao et al. could enhance the oil recovery by 8.0% of OOIP in the presence of 3.0 wt.% NaCl at 80 °C [45]. Ogolo et al. stated that some NPs are suitable for EOR projects, and the presence of ethanol could efficiently improve the EOR performance. Besides, they showed that the mechanisms by which NPs could improve the oil recovery included reduction of oil–water interfacial tension, oil viscosity, and mobility ratio, and modification of the rock surface wettability and the porous medium permeability [46]. Rezaei et al. studied combinations of various surfactants (namely, cocamidopropyl betaine (CAPB), sodium dodecylbenzene sulfonate (SDBS), coconut di-ethanol amide (CDEA), and linear alkylbenzene sulfonic acid (LASA)), alkalis (i.e., sodium carbonate and sodium tetraborate), and nanoparticles (viz., silica, zinc oxide, and cloisite 30B) for EOR from dolomite rock sample [47]. They could achieve almost 19.7% and 12.2% enhancement in ultimate oil recovery from dolomite rock samples by CAPB+sodium carbonate and CAPB+silica nanoparticle, respectively. Suleimanov et al. could achieve up to 18% more oil recovery, in comparison with the surfactant flooding, by combining NPs with the surfactant solution. They attributed this oil recovery improvement to the decrease in the interfacial tension of the nanoparticle-containing system [48]. Handraningat et al. studied the effect of NPs concentration on the efficiency of the nano-EOR flooding and could reduce the probability of pore blockage of sandstone rocks by applying an optimum range for the concentration of NPs [49]. Despite the widespread use of nanotechnology in the EOR schemes, there is still little knowledge on the application of the surfactant–NP composites as a new trend in the field of chemical EOR.

In this study, firstly, the unsteady-state oil–water relative permeability curves of two core plugs with similar flow zone indices (FZI) but different PSDs were compared during synthetic formation water (SFW) flooding, as a secondary EOR technique. Afterward, an extended series of experiments, including the IFT tests between the aqueous phases containing the alpha olefin sulfonate (AOS) surfactant + silica NPs and the oil phase, and the contact angle measurements, were carried out to determine a promising mixture of the surfactant and the NPs for EOR purposes. Zeta potential measurements and dynamic light scattering (DLS) analysis were also performed to evaluate the stability of the silica NPs in the presence of the anionic surfactant. Finally, as a first-time investigation into the efficiency of the surfactant + NP flooding considering the effect of PSDs, core flooding experiments were done on core plugs of the same rock type but different pore throat size distributions.

2. Materials and Methods

In the following section, materials used in this study and their information are listed. An introduction to experimental procedures and the setups used for tests are also covered in this segment.

2.1. Materials

2.1.1. Surfactant

Alpha olefin sulfonate (AOS) in solid white powders is an anionic surfactant that was used in this study. Figure 1 presents the chemical structure of the AOS surfactant.

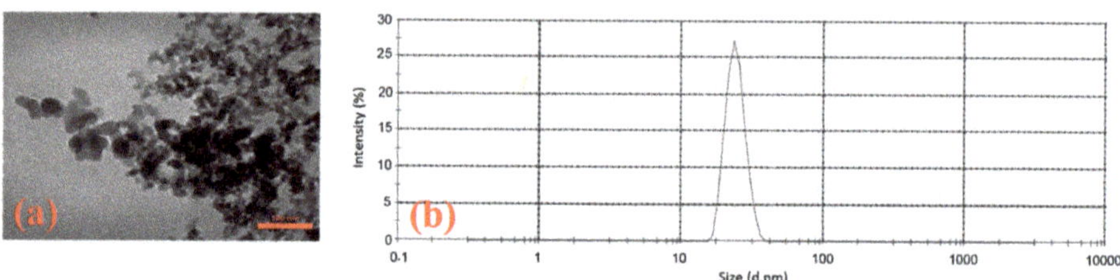

Figure 1. Chemical structure of alfa olefin sulfonate.

2.1.2. Nanoparticle

Hydrophilic silica nanoparticles (non-porous, 25 nm, the specific surface area of 200 m^2/g, the density of 2.4 g/cm^3, and purity +99.5%), which were purchased from Sigma-Aldrich Company, Taufkirchen, Germany were used in experiments. Transmission electron microscopy (TEM) analysis on the silica powders was performed using a transmission electron microscope (Model: Philips EM208S 100KV, Nicosia, Cyprus). In order to investigate the size distribution of silica NPs, a dynamic light scattering (DLS) test was performed using the Malven ZS Nano analyzer (Malven Instrument Inc., London, UK). Results of TEM and DLS analysis were depicted in Figure 2a,b, respectively. As the results of the DLS test revealed, the size distribution of silica NPs is 18 nm to 38 nm, with an average size of 25 nm. According to the thin pore throats of the carbonate rocks, determination of the maximum radius of flocculated NPs is a crucial matter in NPs flooding. As determined by the DLS analysis, the maximum flocculated size of silica NPs is almost 38 nm, which confirms the ability of NPs to pass through the pore throats (minimum size of 0.05 μm to 2 μm as measured by mercury injection capillary pressure (MICP) tests).

Figure 2. (**a**) TEM image and (**b**) DLS analysis of the nanoparticle.

2.1.3. Synthetic Brine

A solution of 180 g of sodium chloride (supplied by Merck, Darmstadt, Germany) in one liter of deionized water (DIW) was used as the synthetic brine. The equivalent amount of NaCl is determined for making synthetic brine with consideration of equal ionic strengths for the synthetic and the formation brine.

2.1.4. Crude Oil

The crude oil obtained from one of the Iranian oil fields with the viscosity of 15 cp and the gravity of 31.5° API was used for aging processes and experimental sections. SARA analysis was performed for better characterization of studied crude oil. Separation of saturates, aromatics, resins, and asphaltenes was performed as follows: initially, asphaltene fraction of the crude oil was extracted according to the IP-143 procedure described in detail elsewhere [50]. Briefly, 1 g of the crude oil was mixed with 30 mL of n-heptane, and refluxed for 60 min and placed in a dark room for about 12 h. Asphaltenes was filtrated by filter paper and washed by n-heptane until the complete removal of impurities. Subsequently, the filter paper was washed by toluene. Finally, solvents were evaporated, and asphaltenes were obtained. The soluble fraction of crude oil in n-heptane was further separated to saturate, aromatic, and resin by silica gel packed column chromatography. Saturate, aromatic, and resin fractions were separated by elution of n-hexane, toluene, and toluene/methanol (90/10), respectively. Figures 3 and 4 show the results of SARA analysis and a schematic view of the asphaltene extraction method, respectively.

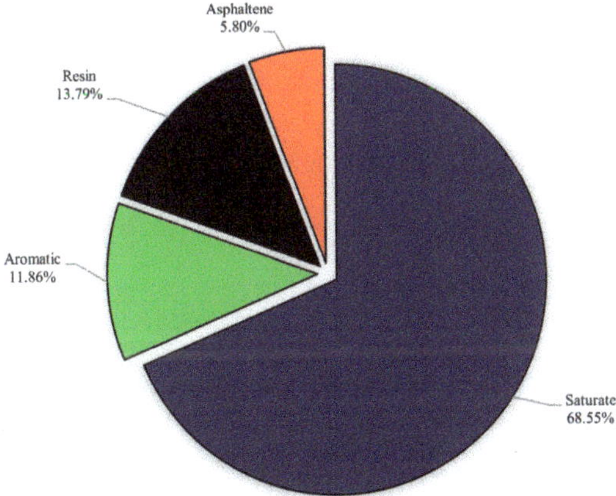

Figure 3. The results of SARA analysis for used crude oil.

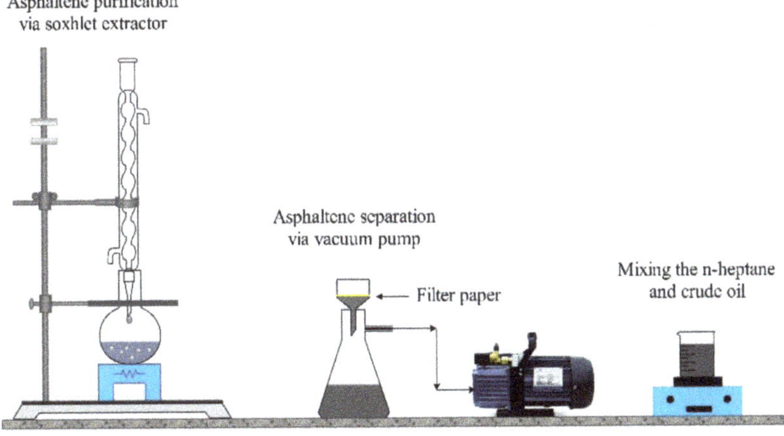

Figure 4. Schematic of the asphaltene extraction process.

2.1.5. Rock Samples

Carbonate rock samples used for core flooding experiments were selected from one of the Iranian oil reservoirs. Results of X-ray diffraction (XRD) analysis on the powdered rock samples are depicted in Figure 5.

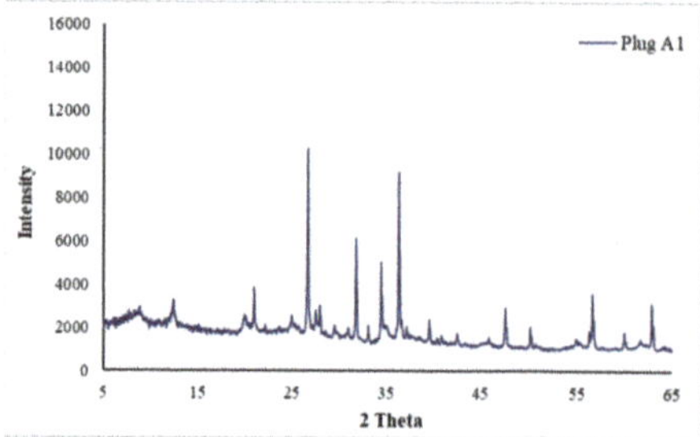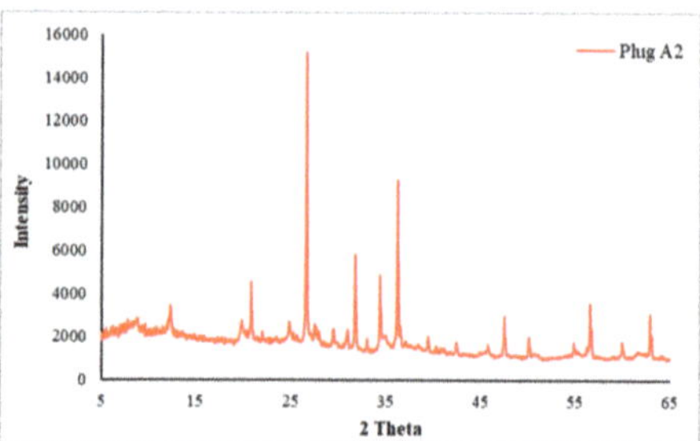

Figure 5. Results of XRD analysis on the powdered A1 (left) and A2 (right) rock samples.

Routine specifications of the core plugs are listed in Table 1.

Table 1. The specifications of the core samples.

Sample	Depth (m)	Length (mm)	Diameter (mm)	Grain Density (g/cm^3)	Porosity (%)	Gas Permeability (mD)	S_{wi} (%)
A1	3609.42	48.12	37.43	2.70	15.73	1.51	22.58
A2	3606.39	47.50	37.51	2.70	13.48	1.05	14.23

2.2. Experimental Procedures

2.2.1. Rock Typing

In order to select representative rock samples between more than 20 core plugs, the common, inexpensive, and efficient flow zone index (FZI) method was applied, and two core plugs with similar rock types were picked. FZI is a practical and well-known method in reservoir engineering to simply identify rock types [51–53]. Results and the method for FZI calculations are described in Section 3.1.

2.2.2. Zeta Potential Measurement

To inquire about the effect of AOS surfactant on the stability of NPs suspensions, two solutions (a) 0.01 wt.% of silica NPs and (b) 0.01 wt.% of AOS + 0.01 wt.% of silica NPs, both in DIW were prepared, and the zeta potential for the two solutions was measured. It is noteworthy that the reported zeta potential values for each solution are the average of three times measurements.

2.2.3. IFT Measurements

A series of IFT measurements were performed to find the optimum concentrations for AOS surfactant and silica NPs in the EOR solution. Drop shape analyzer (DSA 100, KRÜSS, Hamburg, Germany) apparatus was applied in both IFT measurements and determination of the oil droplets contact angle on the oil-wet rock sections. For the purpose of IFT determination between the oil and the aqueous phases, oil droplets were injected into the aqueous phase (as the bulk phase, contains surfactant and NPs) via a syringe; and the image of the oil droplet as it was about to be apart was used for IFT determinations. In-house software was applied for calculating the value of IFTs.

2.2.4. Contact Angle Measurements

Contact angles of the sessile oil droplets on the rock surface were measured via DSA 100 apparatus. In all experiments, the aqueous phase composed of different concentrations of AOS + SiO$_2$ NPs in SFW was used as the bulk fluid. Figure 6 demonstrates a schematic of the equipment used for IFT and contact angle measurements.

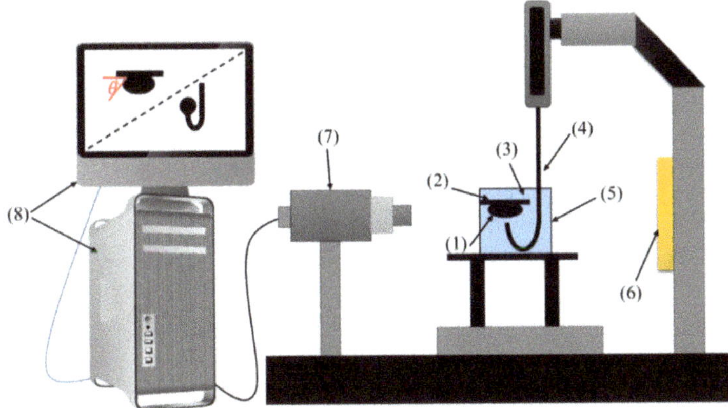

Figure 6. Schematic of the setup for interfacial tension IFT) and contact angle tests: (1) oil droplet, (2) rock section, (3) aqueous solution, (4) needle, (5) cell, (6) light source, (7) CCD camera, (8) processor.

2.2.5. MICP Tests

The distribution of pressure-volume relationships can be explored through mercury injection and monitoring the entry of mercury into a pore system. As the mercury is the non-wetting phase, it should overcome the capillary forces to go into the pore volumes, and the intrusion process will only progress after applying an ever-increasing pressure. A Micromeritics Autopore IV 9500 Porosimeter apparatus, GA, USA (see Figure 7) was used for MICP measurements.

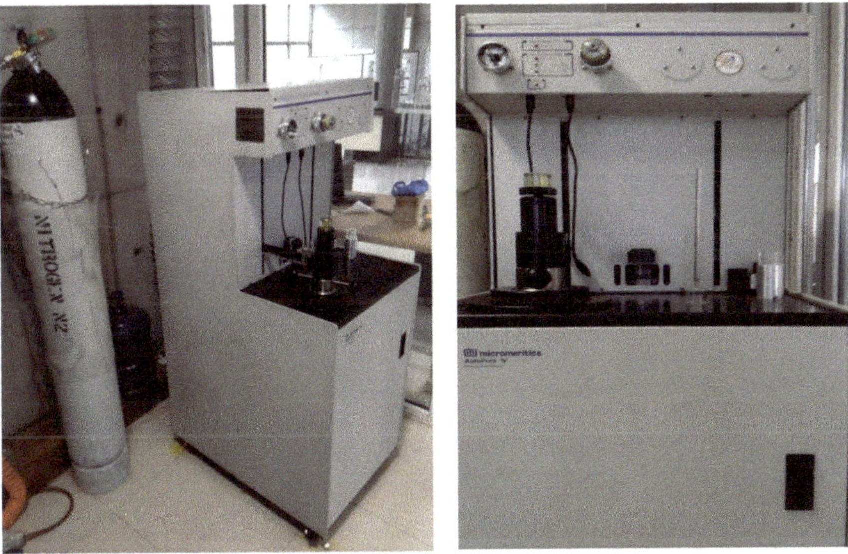

Figure 7. Setup used for mercury injection capillary pressure (MICP) tests.

Following is the procedure of MICP tests applied in this study to determining PSDs in core cuttings:

- Weighting the cleaned and dried sample.
- Selecting a proper penetrometer according to the pore volume of the core cutting.
- Weighting the penetrometer containing the cutting.

- Loading the penetrometer to the low-pressure chamber.
- Measuring the bulk volume of the cutting.
- Increasing the pressure of mercury injection from 0.5 to 30.0 psig incrementally and monitoring the amount of mercury intrusion at each pressure step.
- Loading the sample into the high-pressure chamber.
- Injecting the mercury into the core cutting and raising the pressure (up to 60000 psia) incrementally.
- After equilibrium is established in the last step, the injection pressure is reduced to atmospheric pressure.
- Mercury saturations are calculated as a percentage of the pore volume at each pressure, and the pore volume used for calculation of mercury saturation obtained from the maximum intrusion volume.

2.2.6. Core Flood Experiments

Applicability of the surfactant–NPs solution as an EOR method was examined through displacement tests, and the efficiency of this EOR method in different PSDs was investigated. The core plug is placed into a hassler-type stainless steel core holder, and overburden pressure about 1000 Psi was applied using a manual confining pump. The liquid accumulator and EOR solution cylinder connected to the core holder and a precise pressure transmitter (Model: Rosemount 3051CD, MN, USA) was installed to measure the differential pressure between the inlet and the outlet of the core sample. The HPLC pump injects the distilled water to the transfer vessels and pushes the piston up, which causes the fluids (i.e., SFW, surfactant–nanoparticle solutions, or the crude oil) to be injected into the core plug by a specific rate. The outlet pressure in all core flooding experiments was ambient pressure. Eventually, produced fluid was collected in a calibrated burette. Figure 8 depicts a diagram of the setup used for core flooding experiments.

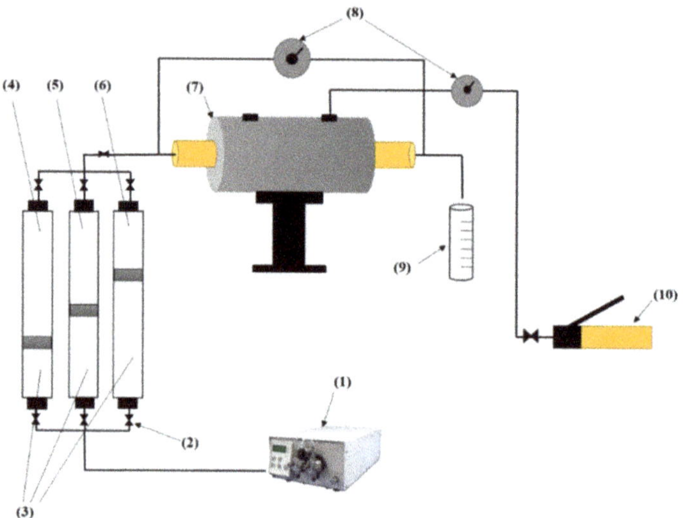

Figure 8. Diagram of the setup used for core flooding tests: (1) high-pressure liquid chromatographic (HPLC) pump, (2) valve, (3) distilled water, (4) synthetic formation water, (5) dead oil, (6) surfactant/nanoparticle solution, (7) core holder, (8) differential pressure transmitter, (9) calibrated burette, (10) overburden pressure.

3. Results and Discussion

3.1. Core Sample Selection

FZI rock typing technique was applied on more than 20 core plugs cached from the depths of 3604 m to 3610 m of the reservoir, and the two core plugs which were principally alike were selected. This method includes the three following equations (see Equations (1)–(3)) [51]:

$$RQI = 0.0314 \sqrt{(k/\varphi)}, \tag{1}$$

$$\varphi_z = \varphi/(1 - \varphi), \tag{2}$$

$$FZI = RQI/\varphi_z, \tag{3}$$

where RQI is the rock quality index (μm), K is the permeability of the core plug (mD), φ is the effective porosity (fraction), φ_z is the normalized amount of the effective porosity, and the ratio of RQI to normalized porosity is called the flow zone index (μm).

The calculated FZI for cores A1 and A2 were 0.52 and 0.56, respectively, which indicates that the two samples have similar FZI; thus, they are classified into the same rock type or hydraulic flow unit (HFU) [51,54,55].

3.2. Static Stability of the Solution

In order to investigate the stability of NPs into the aqueous solutions, zeta potential measurements were conducted on two pre-prepared suspensions, including (a) 0.01 wt.% of SiO_2 and (b) 0.01 wt.% of SiO_2 + 0.01 wt.% of AOS surfactant, in DIW. Results of zeta potential measurements for the NPs and the AOS–NPs are shown in Figure 9. A negative zeta potential value signifies that the forces between the silica NPs and the AOS surfactant monomers used in this study are repulsive, which causes the nanoparticles not to be accumulated [56]. Investigating colloid stability is one of the most common applications of zeta potential data. Figure 10 shows guidelines classifying NPs-dispersions with different ranges of zeta potential values [56,57].

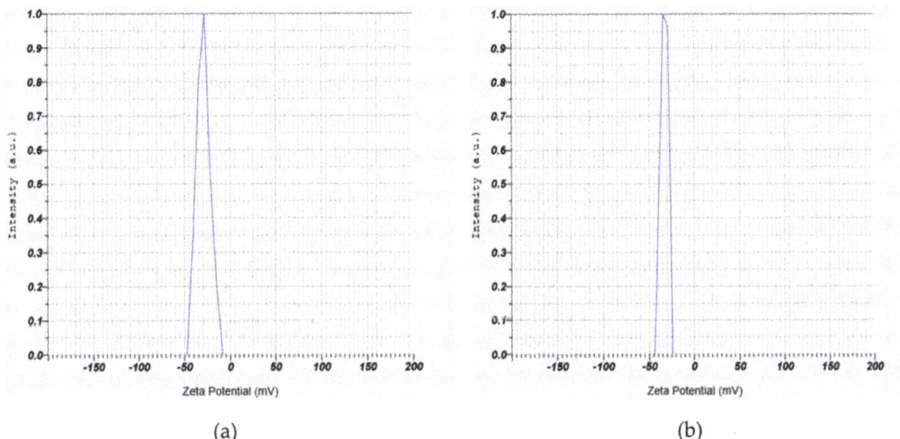

(a) (b)

Figure 9. Zeta potential measurements for (**a**) SiO_2 and (**b**) alpha olefin sulfonate (AOS) + SiO_2 solutions.

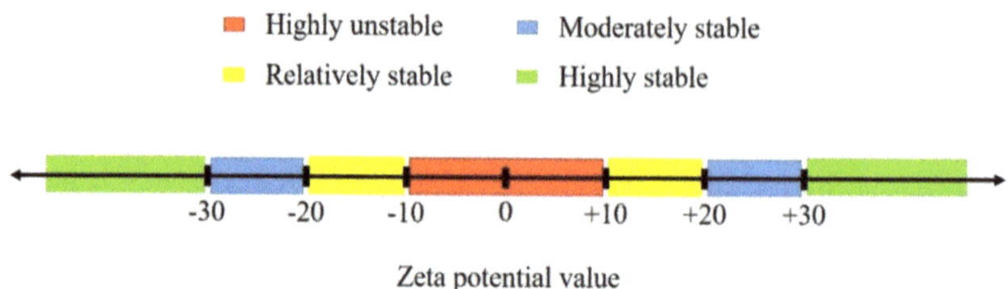

Figure 10. Different ranges of zeta potential values and the level of stability of nanoparticles into the aqueous phase.

The average amounts of zeta potential measurements were reported −29.8 and −32.5 for SiO_2 and AOS + SiO_2 solutions, respectively. As can be deduced from Figure 10, the results of zeta potential determinations depicted that the presence of the anionic surfactant could efficiently improve the stability of silica NPs into the solution.

In order to investigate the size distribution of the SiO_2 NPs, separately and in combination with the AOS surfactant, two aqueous solutions containing (a) SiO_2 (0.005 wt.%) and (b) AOS (0.005 wt.%) + SiO_2 (0.005 wt.%) were prepared and the DLS analysis was performed on both. The results of DLS tests demonstrated that the radius of aggregated NPs as 178 and 146 nm for the solutions (a) and (b), respectively. The smaller radius of the aggregated NPs for the AOS + SiO_2 solution indicates the lower inclination of the NPs to sedimentation.

3.3. IFT between the Crude Oil and Aqueous Solutions

Two series of IFT measurements were conducted to finally determine the optimum concentrations for the AOS surfactant and the silica NPs in aqueous solution. The first series of oil–water IFT measurements were performed, and the critical micelle concentration (CMC) for the AOS surfactant was measured as 0.067 wt.%. Figure 11 shows the results of IFT measurements between different concentrations of the AOS surfactant in SFW and crude oil.

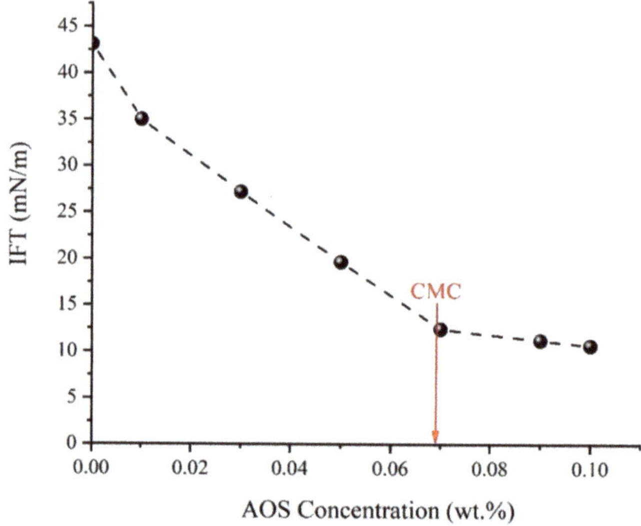

Figure 11. Crude oil–aqueous solution IFT at different concentrations of AOS surfactant.

In the next series of experiments, constant concentrations of 0.06, 0.07, and 0.08 wt.% of AOS surfactant and 0.05, 0.1, and 0.2 wt.% of SiO_2 NPs were added to SFW, and the oil–aqueous phase IFT values were determined. As can be concluded from Figure 12, the addition of silica NPs to the surfactant solution causes a further reduction in IFT values. In these concentrations for silica NPs, the optimum concentration for the presence of NPs in the aqueous solution was determined as 0.1 wt.%.

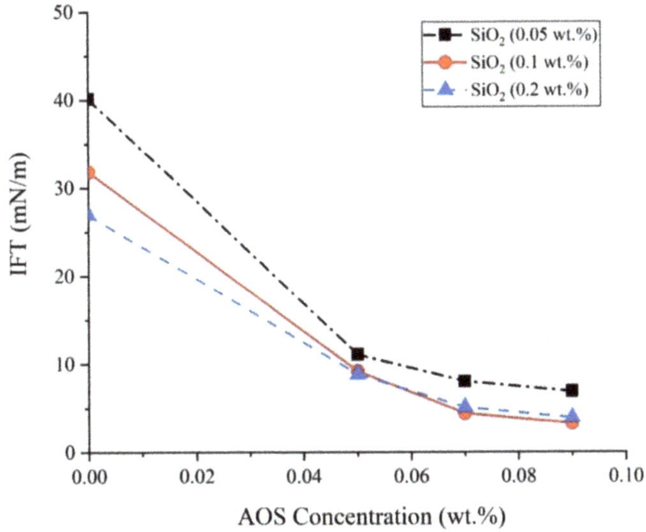

Figure 12. IFT values between the crude oil and the aqueous solution.

3.4. Wettability Measurement

In this section, we measured the variation of the contact angle of oil droplets on oil-wet rock sections with the aid of a sessile drop technique. For this purpose, the restored rock sections were used to measure the wettability alterations by soaking them into the SFW and the EOR solution. The contact angle of oil droplets on rock sections was determined (a) before aging, (b) after aging, (c), and (d) after 12 and 24 h soaking into the aqueous phases, respectively. Figure 13 shows the results of contact angle measurements at different times when the SFW and the EOR solution (0.067 wt.% of AOS + 0.1 wt.% SiO_2) are the bulk phases.

The disjoining pressure defined as the pressure required to overcome the fluid's adhesion force to the solid surface to remove the liquid from the surface. In this case, the disjoining pressure is defined as the attractive interactions between the aqueous phase and the oil film. Through this mechanism, a wedge film of the discontinuous oil phase attached to the rock surface is created by the NPs, which were dispersed into the aqueous phase (see Figure 14). As a specific point of view, the formation of these wedge films, which are considerably influenced by Brownian motion and electrostatic repulsive forces between nanoparticles, is the primary mechanism for wettability alteration of the rock sections from oil-wet to water-wetness [58]. On the other hand, by increasing the concentration of nanoparticles, the repulsive forces between the NPs, and also the intensification of these repulsive forces due to the presence of surfactant monomers with the same electrical charge, increases the disjoining pressure and Brownian motion. So, the performance of the aqueous phase in decreasing the contact angle values improves as the concentration of nanoparticles increases [58–60]. However, the point to note here is that increasing the concentration of NPs, in addition to raising the costs, enhances the possibility of NPs deposition, which leads to formation damage due to plugging the rock pores and pore throats [61,62].

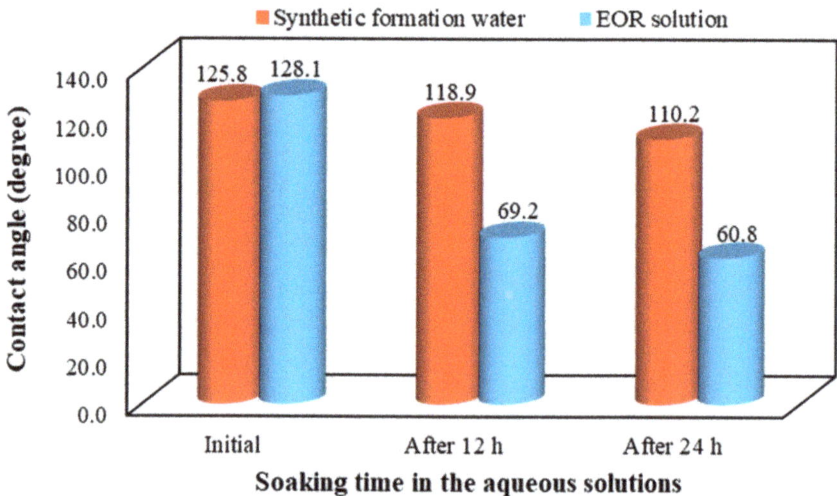

Figure 13. Effects of soaking time of the rock sections in red: synthetic formation water and blue: the enhanced oil recovery (EOR) solution on the oil-rock contact angle.

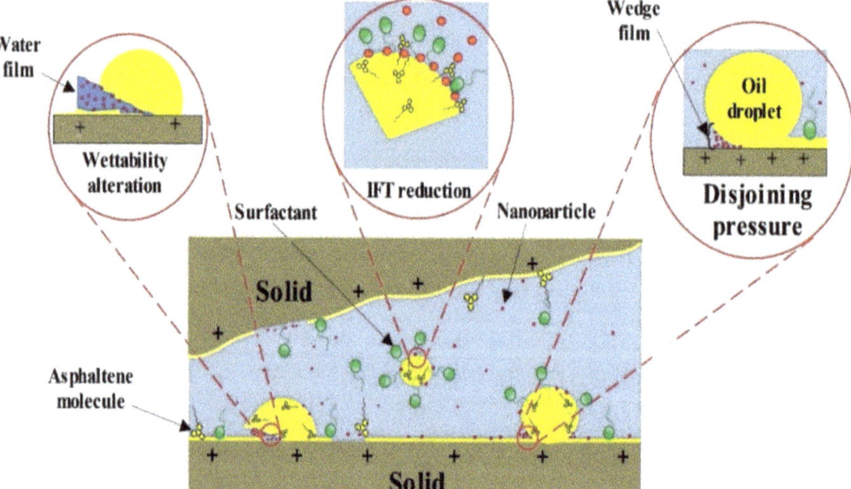

Figure 14. Schematic view of surface interactions of the carbonate rock–oil–aqueous phase (containing surfactant + NPs) system.

To have a better comparison of the ability of different aqueous solutions in altering the rock surface wettability, the effect of the initial value of the contact angle was eliminated. Thus, the wettability alteration index (WAI) was determined using Equation (4) [63], and the results are shown in Figure 15.

$$WAI = (\theta_0 - \theta_f)/(\theta_0 - \theta_i), \tag{4}$$

where θ_0 and θ_i are contact angles after and before aging, respectively, and θ_i is contact angle after being in contact with the EOR solution. As deduced from Equation (4); the closer the value of WAI to 1, the more the wettability changes due to the aqueous bulk solution. Figure 15 also illustrates the superiority of the EOR solution performance in changing the wettability of rock surfaces towards water-wetness in comparison with the SFW.

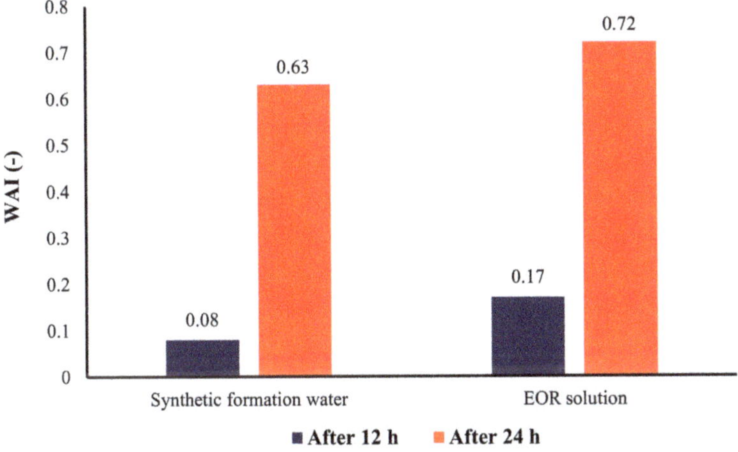

Figure 15. Effects of soaking time in different aqueous solutions on the wettability alteration index (WAI) of the carbonate rock sections.

3.5. Determination of the PSD

The fraction of the pore volume injected against pore throat radius is represented through the data of MICP tests. As shown in Equation (5), the differential of pore volume injected provides a function for pore throat size distribution.

$$PSD = dv/dlog(r), \tag{5}$$

The differential equation will be calculated numerically. The central difference method (Equation (6)) is used to calculate PSD as:

$$PSD_i = (V_{i+1} - V_{i-1})/(\log(r_{i+1}) - \log(r_{i-1})), \tag{6}$$

Using Equation (7), the PSDs of the core plugs were smoothed and then normalized to 1 by applying the Equation (8):

$$PSD_i = (PSD_{i-1} + 2PSD_i + PSD_{i+1})/4, \tag{7}$$

$$PSD_{normal,i} = PSD_i/PSD_{max}, \tag{8}$$

The normalized PSD is represented in graphical shape along with the saturation of mercury against pore throat radius. Figures 16 and 17 show the PSDs determined through MICP tests for the core plugs A1 and A2, respectively.

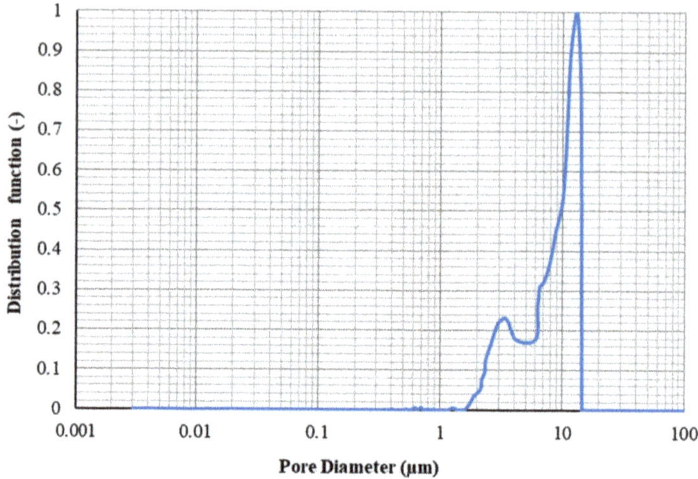

Figure 16. Pore throat size distribution for core A1.

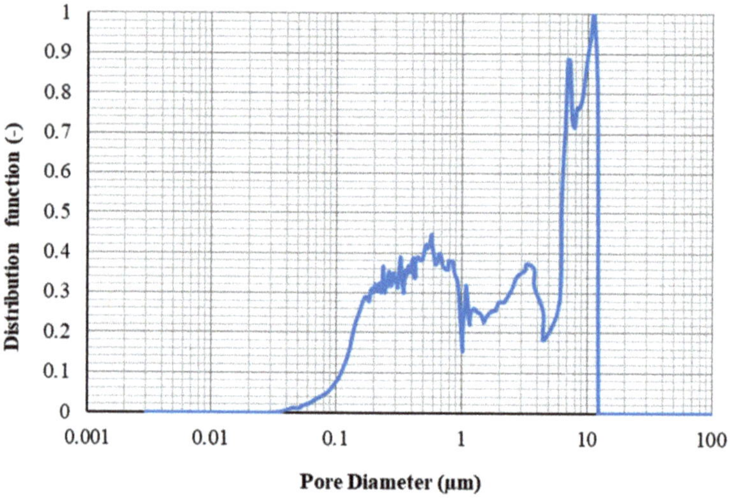

Figure 17. Pore throat size distribution for core A2.

As can be seen from Figure 17, the PSDs for core A2 cover a broad range of pore throat sizes from 0.03 to 14 µm uniformly, and the average diameter of the pores is 763.9 nm. Although the range of pore throat diameter for sample A1 (from 1.8 to 15 µm- see Figure 16) is comparable to that for sample A2, the average pore diameters for this rock sample (7609.5 nm) is far greater than that for sample A1. Therefore, this can be concluded that for core sample A1, there are larger pore throat sizes in it, which may not be connected effectively, and core sample A2, pores would have better connections together because of the extensive range of pore throat radiuses.

The total pore areas for the core samples A1 and A2 were also estimated as 0.273 and 0.035 (m^2/g), respectively.

3.6. Flooding Experiments

In order to investigate the effect of pore throat size distributions on the flowing behavior of the core samples, two series of flooding experiments were conducted on each core plug. Figure 18 depicts the core preparation procedure carried out before unsteady-state relative permeability measurements.

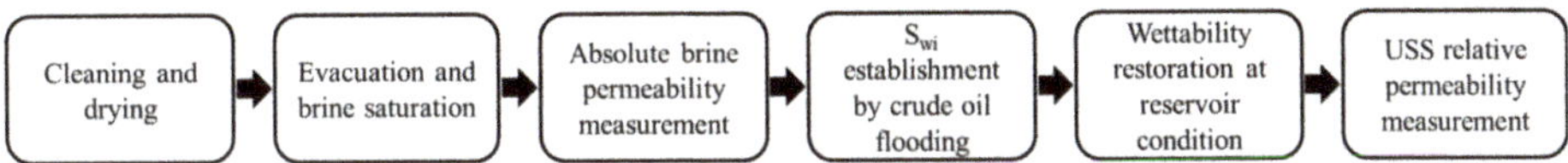

Figure 18. Steps to core sample preparation.

Core plugs were immersed in crude oil at 90 °C and 3500 psi for 21 days to restore the wettability of core plugs to reservoir conditions.

In the interest of investigating the effect of different pore throat size distributions on the flowing behavior of the oil and water phase, two-phase unsteady-state relative permeability curves were established based on the data of oil production and pressure drop across the core plug during SFW flooding.

After loading the core plug into the core holder, crude oil was injected into the sample to (1) replace the oil into the core plug with fresh oil and (2) to obtain oil permeability at irreducible water saturation (S_{wi}). Figure 19 indicates the relative permeability curves for core samples A1 and A2. As depicted in Figure 19, the amount of oil recovered by water flooding increases as the homogeneity of the pore throat size and pore connectivity increase. The cross point of water and oil relative permeability curves for core plugs A1 and A2 were at water saturation of 32% and 36%, respectively. It meant that, although the aging process could not sufficiently change the wettability of rock samples to oil wetness, it was more efficient in core sample A2, because of the better connections between pores.

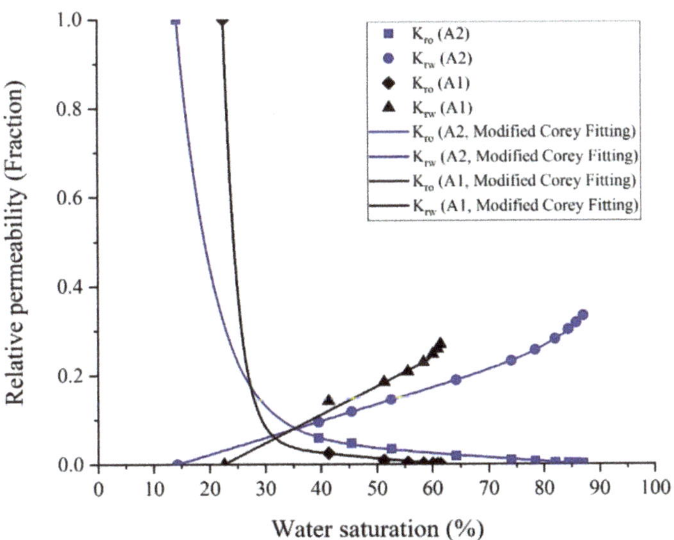

Figure 19. Relative permeability curve for cores A1 and A2.

The residual oil saturation for the core sample A2 is about 25% less than the core A1. Unlike what Wardlaw and Magara [6,7] stated theoretically, the higher oil recovery for a rock sample obtained from core flooding experiments could not exclusively be the result of having a higher absolute gas permeability or larger pore sizes, and the PSDs and the homogeneity of pore sizes are of particular importance.

To compare the ability of the pore network of the core samples in oil transmission, the relative permeability ratio of water and oil (K_{rw}/K_{ro}) in a similar range of water saturation (S_w) for core sample A1 and A2 are represented in Figure 20. As demonstrated in Figure 20, the ratio of water to oil relative permeability data for core sample A2 is much lower than that for core sample A1, which implies the higher ability of the core sample A2 in oil transmissibility than that for core sample A1. Furthermore, it can be concluded from Figure 20 that in similar aging conditions, the core A1 showed more oil-wet behavior, which this behavior can be attributed to its large pore size distribution and more homogeneous structure.

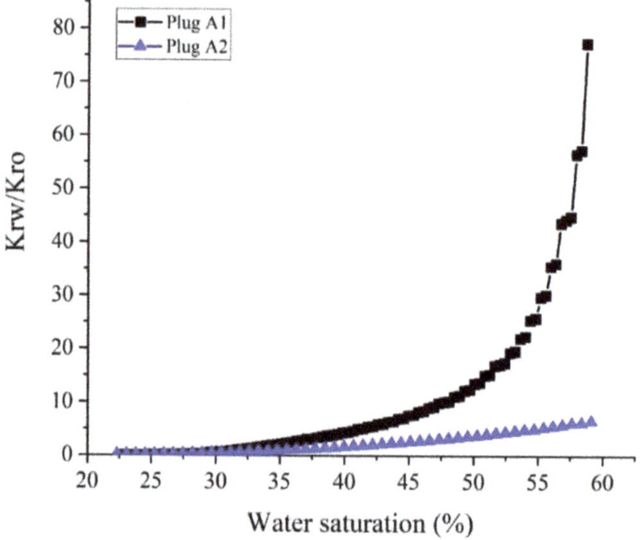

Figure 20. The ratio of water to oil relative permeability (K_{rw}/K_{ro}) for core samples A1 and A2 at overlapped water saturations.

The experimental procedure for measuring the PSD and the USS relative permeability are available in Supplementary Materials (see Section S.1 in Supplementary file and Figures S1 to S3 and Tables S1 to S3).

3.7. Enhanced Oil Recovery Tests

After flooding with formation water, a solution of SFW containing 0.1 wt.% AOS surfactant + 0.1 wt.% silica NPs was flooded into the core samples with a constant flow rate of 0.02 cc/min. Figure 21 shows the amount of oil recovered after secondary and tertiary modes of displacement tests performed on the core plugs.

For sample A2, most of the oil is recovered through secondary water flooding. This is because of the homogeneity of the pore throat sizes in this core plug, which causes that the aqueous phase sweeps the oil easily out of the pores, and reaches the residual oil saturation after water flooding to a very low value. However, in sample A1, a large amount of oil remained unswept after flooding with SFW. Early water breakthrough time and also the high amount of unswept oil remained within the core plug after water flooding confirmed the presence of some large pore throats into the core plug, which drives the aqueous phase flow to the end of the plug.

Distribution of some large pores alongside a negligible range of small pores in the structure of sample A1 increases the probability of having a very high aspect ratio in the pore network of the core plug. Therefore, the significant amount of oil left behind into the core plug after secondary core flooding tests makes this rock sample an appropriate candidate for EOR plans. The best way to recover the unswept oil remaining in the core plug after water flooding is to alter the rock surface wettability towards water-wetness and reduce oil–water IFT; thus, oil transmissibility into the porous media will increase. NPs are able to move through the small and inaccessible pores and can alter the wettability of the surface of the pores to water-wetness. The addition of the AOS surfactant to the NPs suspension will increase the stability of the NPs into the aqueous phase and also reduces the IFT between the remained oil into the pores and the aqueous phase, and subsequently, the residual oil, which is remained unswept can be recovered. Application of surfactant + NPs as the EOR technique could enhance oil recovery from core plugs A1 and A2 up to about 2.5% and 8.6%, respectively.

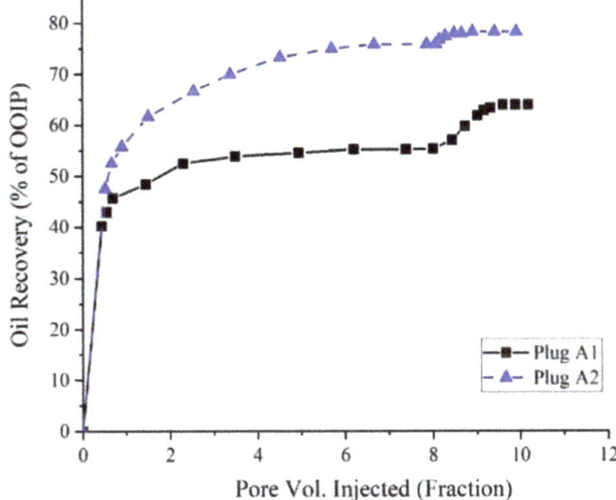

Figure 21. Oil recovered from the secondary and tertiary oil recovery processes.

4. Conclusions

In the present study, for the first time, the simultaneous effects of PSDs and a chemical-based EOR method were investigated experimentally through an extended series of experiments, and the following conclusions are obtained:

- The presence of 0.1 wt.% silica NPs in combination with the AOS surfactant (CMC), reduces the IFT value from 12.5 mN/m to 4.4 mN/m. The results of zeta potential and DLS tests confirmed the synergy of AOS surfactant and silica nanoparticles.
- Results of contact angle measurements, as well as oil–water relative permeability tests, indicate the capability of the AOS + SiO_2 solution in altering the wettability of carbonate rocks toward water-wetness.
- For core sample A2 with a wide range of uniform pore sizes, the amount of oil recovered by secondary water flooding was more than that for core plug A1. Thus, in conventional water flooding, the homogeneity of the pore sizes is more important than the dimensions of the pores.
- Based upon the results of core flooding experiments, which showed more oil recovery in water flooding for the plug with homogeneous PSDs, it can be deduced that using the results of MICP tests is a more reliable method than the FZI technique for rock typing in carbonate rocks.

- At the same range of water saturation, the ratio of water to oil relative permeability for core plug with better pore connectivity is much lower than that for the core plug with poor pore connections. This shows how the better connection between the pores can influence oil recovery by water flooding, which is in good agreement with the results of core flooding tests.
- Results of displacement tests outlined that applying the surfactant + NPs solution as an EOR technique could recover up to 2.5% and 8.6% more oil from the core plugs with homogeneous and non-homogeneous PSDs, respectively.

Supplementary Materials: The following are available online at http://www.mdpi.com/2079-4991/10/5/972/s1, Figure S1: Desiccator used for vacuum saturation of the core plugs, Figure S2: Schematic view of the setup used for USS water/oil relative permeability measurements, Figure S3: Water/oil relative permeability curves, Table S1: Routine core specifications and conditions of the water/oil relative permeability test, Table S2: The results of USS oil/water relative permeability test, Table S3: Parameters of the Modified Corey Model.

Author Contributions: Conceptualization, A.R., A.H.-S.; methodology, A.R., H.A., Z.D.; software, A.R., Z.D.; validation, A.R., A.H.-S.; formal analysis, A.R., Z.D., H.A.; investigation, A.R., H.A.; resources, A.R., Z.D.; writing—original draft preparation, A.R.; writing—review and editing, A.H.-S. and A.M.; visualization, A.R., N.N.; supervision, A.M., A.H.-S. Funding, A.M.; validation, A.M. All authors have read and agreed to the published version of the manuscript.

Funding: This research received no external funding.

Acknowledgments: The authors would like to acknowledge the support and assistance of the Abdal Industrial Project Management (MAPSA) Company during this study. In addition, we acknowledge the support of this work by the Hungarian State and the European Union under the EFOP-3.6.1-16-2016-00010 project and the 2017-1.3.1-VKE-2017-00025 project.

Conflicts of Interest: The authors declare no conflicts of interest.

Abbreviations

AOS	Alpha olefin sulfonate
DIW	Deionized water
DLS	Dynamic light scattering
EOR	Enhanced oil recovery
FZI	Flow zone index
IFT	Interfacial tension
MICP	Mercury injection capillary pressure
NP	Nanoparticle
PSD	Pore throat size distribution
SFW	Synthetic formation water
TEM	Transmission electron microscopy
WAI	Wettability alteration index
XRD	X-ray diffraction

References

1. Al-Shalabi, E.W.; Ghosh, B. Effect of pore-scale heterogeneity and capillary-viscous fingering on commingled waterflood oil recovery in stratified porous media. *J. Pet. Eng.* **2016**, *2016*, 1708929. [CrossRef]

2. Rezaei, A.; Riazi, M.; Escrochi, M. Investment Opportunities in Iranian EOR Projects. *Saint Petersbg.* **2018**, *2018*, 1–5.

3. Yazdanpanah, A.; Rezaei, A.; Mahdiyar, H.; Kalantarials, A. Development of an efficient hybrid GA-PSO approach applicable for well placement optimization. *Adv Geo-Energy Res.* **2019**, *3*, 365–374. [CrossRef]

4. Serag El Din, S.; Dernaika, M.; Hannon, L.; Kalam, Z. The Effect of Rock Properties on Remaining and Residual Oil Saturation in Heterogeneous Carbonate Rocks. In Proceedings of the SPE Middle East Oil and Gas Show and Conference, Manama, Bahrain, 10–13 March 2013; Society of Petroleum Engineers: Richardson, TX, USA.

5. Aliakbardoust, E.; Rahimpour-Bonab, H. Integration of rock typing methods for carbonate reservoir characterization. *J. Geophys Eng* **2013**, *10*, 55004. [CrossRef]

6. Wardlaw, N.C.; Cassan, J.P. Estimation of recovery efficiency by visual observation of pore systems in reservoir rocks. *Bull. Can. Pet. Geol* **1978**, *26*, 572–585.

7. Magara, K. Estimation of Recovery Efficiency by Visual Observation of Pore Systems in Reservoir Rocks: Discussion. *Bull. Can. Pet. Geol* **1979**, *27*, 400–401.

8. Chatzis, I.; Morrow, N.R.; Lim, H.T. Magnitude and detailed structure of residual oil saturation. *Soc Pet. Eng J.* **1983**, *23*, 311–326. [CrossRef]

9. Elgaghah, S. A Novel Technique for the Determination of Microscopic Pore Size Distribution of Heterogemeous Reservoir Rocks. In Proceedings of the Asia Pacific Oil and Gas Conference and Exhibition, Jakarta, Indonesia, 30 October–1 November 2007; Society of Petroleum Engineers: Richardson, TX, USA.

10. Kianinejad, A.; DiCarlo, D.A. Three-phase oil relative permeability in water-wet media: A comprehensive study. *Transp. Porous Media* **2016**, *112*, 665–687. [CrossRef]

11. Medina, C.R. Influence of Porosity, Permeability, and Pore Size Distribution on Storability, Injectivity, and Seal Efficiency of Carbonate Reservoirs and Shale Caprock: A Multi-Technique Approach for Geologic Carbon Sequestration. Ph.D. Thesis, Indiana University, Bloomington, IN, USA, 2019.

12. Gharibshahi, R.; Jafari, A.; Haghtalab, A.; Karambeigi, M.S. Application of CFD to evaluate the pore morphology effect on nanofluid flooding for enhanced oil recovery. *RSC Adv.* **2015**, *5*, 28938–28949. [CrossRef]

13. Weger, R.J.; Eberli, G.P.; Baechle, G.T.; Massaferro, J.L.; Sun, Y.-F. Quantification of pore structure and its effect on sonic velocity and permeability in carbonates. *Am. Assoc. Pet. Geol. Bull.* **2009**, *93*, 1297–1317. [CrossRef]

14. Rangel-German, E.R.; Kovscek, A.R. Experimental and analytical study of multidimensional imbibition in fractured porous media. *J. Pet. Sci. Eng.* **2002**, *36*, 45–60. [CrossRef]

15. Nwidee, L.N.; Lebedev, M.; Barifcani, A.; Sarmadivaleh, M.; Iglauer, S. Wettability alteration of oil-wet limestone using surfactant-nanoparticle formulation. *J. Colloid Interface Sci.* **2017**, *504*, 334–345. [CrossRef] [PubMed]

16. Ghadami, N.; Rasaei, M.R.; Hejri, S.; Sajedian, A.; Afsari, K. Consistent porosity–permeability modeling, reservoir rock typing and hydraulic flow unitization in a giant carbonate reservoir. *J. Pet. Sci. Eng.* **2015**, *131*, 58–69. [CrossRef]

17. Porras, J.C. Determination of rock types from pore throat radius and bulk volume water, and their relations to lithofacies, Carito Norte field, eastern Venezuela basin. In Proceedings of the SPWLA 39th Annual Logging Symposium, Keystone, CO, USA, 26–28 May 1998; Society of Petrophysicists and Well-Log Analysts: Houston, TX, USA.

18. Leal, L.; Barbato, R.; Quaglia, A.; Porras, J.C.; Lazarde, H. Bimodal behavior of mercury-injection capillary pressure curve and its relationship to pore geometry, rock-quality and production performance in a laminated and heterogeneous reservoir. In Proceedings of the SPE Latin American and Caribbean Petroleum Engineering Conference, Buenos Aires, Argentina, 25–28 March 2001; Society of Petroleum Engineers: Richardson, TX, USA.

19. Skalinski, M.; Jeroen Kenter, S.P.E.; Jenkins, S.; Tengizchevroil, T.T. Updated rock type definition and pore type classification of a carbonate buildup, Tengiz field, Republic of Kazakhstan. *Soc. Pet. Eng.* **2010**, 139986. [CrossRef]

20. Lehmann, C.T.; Mohamed, K.I.; Cobb, D.O.; Al Hendi, A. Rock-typing of upper jurassic (Arab) carbonates offshore Abu Dhabi. In Proceedings of the Abu Dhabi International Petroleum Exhibition and Conference, Abu Dhabi, UAE, 3–6 November 2008; Society of Petroleum Engineers: Richardson, TX, USA.

21. Colombo, F.; del Monte, A.A.; Balossino, P.; Paparozzi, E.; Valdisturlo, A.; Tarchiani, C. MICP-Based Elastic Rock Typing Characterisation of Carbonate Reservoir. In Proceedings of the SPE Europec featured at 80th EAGE Conference and Exhibition, Copenhagen, Denmark, 11–14 June 2018; Society of Petroleum Engineers: Richardson, TX, USA.

22. Green, D.W.; Willhite, G.P. *Enhanced Oil Recovery*; Henry, L. Doherty Memorial Fund of AIME, Society of Petroleum Engineers: Richardson, TX, USA, 1998; Volume 6.

23. Saxena, N.; Kumar, A.; Mandal, A. Adsorption analysis of natural anionic surfactant for enhanced oil recovery: The role of mineralogy, salinity, alkalinity and nanoparticles. *J. Pet. Sci. Eng.* **2019**, *173*, 1264–1283. [CrossRef]

24. Karimi, M.; Al-Maamari, R.S.; Ayatollahi, S.; Mehranbod, N. Wettability alteration and oil recovery by spontaneous imbibition of low salinity brine into carbonates: Impact of Mg^{2+}, $SO_4{}^{2-}$ and cationic surfactant. *J. Pet. Sci. Eng.* **2016**, *147*, 560–569. [CrossRef]
25. Standnes, D.C.; Austad, T. Wettability alteration in chalk: 2. Mechanism for wettability alteration from oil-wet to water-wet using surfactants. *J. Pet. Sci. Eng.* **2000**, *28*, 123–143. [CrossRef]
26. Derikvand, Z.; Riazi, M. Experimental investigation of a novel foam formulation to improve foam quality. *J. Mol. Liq.* **2016**, *224*, 1311–1318. [CrossRef]
27. Yassin, M.R.; Arabloo, M.; Shokrollahi, A.; Mohammadi, A.H. Prediction of surfactant retention in porous media: A robust modeling approach. *J. Dispers Sci. Technol.* **2014**, *35*, 1407–1418. [CrossRef]
28. Li, K.; Jing, X.; He, S.; Wei, B. Static adsorption and retention of viscoelastic surfactant in porous media: EOR implication. *Energy Fuels* **2016**, *30*, 9089–9096. [CrossRef]
29. Levitt, D.; Bourrel, M. Adsorption of EOR Chemicals Under Laboratory and Reservoir Conditions, Part III: Chemical Treatment Methods. In Proceedings of the SPE Improved Oil Recovery Conference, Tulsa, OK, USA, 11–13 April 2016; Society of Petroleum Engineers: Richardson, TX, USA.
30. Wu, Y.; Chen, W.; Dai, C.; Huang, Y.; Li, H.; Zhao, M.; He, L.; Jiao, B. Reducing surfactant adsorption on rock by silica nanoparticles for enhanced oil recovery. *J. Pet. Sci. Eng.* **2017**, *153*, 283–287. [CrossRef]
31. Sofla, S.J.D.; Sharifi, M.; Sarapardeh, A.H. Toward mechanistic understanding of natural surfactant flooding in enhanced oil recovery processes: The role of salinity, surfactant concentration and rock type. *J. Mol. Liq.* **2016**, *222*, 632–639. [CrossRef]
32. Derikvand, Z.; Rezaei, A.; Parsaei, R.; Riazi, M.; Torabi, F. A mechanistic experimental study on the combined effect of Mg^{2+}, Ca^{2+}, and $SO_4$2-ions and a cationic surfactant in improving the surface properties of oil/water/rock system. *Colloids Surfaces A Physicochem Eng. Asp.* **2019**, 124327. [CrossRef]
33. Bi, Z.; Liao, W.; Qi, L. Wettability alteration by CTAB adsorption at surfaces of SiO2 film or silica gel powder and mimic oil recovery. *Appl. Surf. Sci.* **2004**, *221*, 25–31. [CrossRef]
34. Amirianshoja, T.; Junin, R.; Idris, A.K.; Rahmani, O. A comparative study of surfactant adsorption by clay minerals. *J. Pet. Sci. Eng.* **2013**, *101*, 21–27. [CrossRef]
35. Franco, C.A.; Zabala, R.; Cortés, F.B. Nanotechnology applied to the enhancement of oil and gas productivity and recovery of Colombian fields. *J. Pet. Sci. Eng.* **2017**, *157*, 39–55. [CrossRef]
36. Cheraghian, G.; Hendraningrat, L. A review on applications of nanotechnology in the enhanced oil recovery part B: Effects of nanoparticles on flooding. *Int. Nano Lett.* **2016**, *6*, 1–10. [CrossRef]
37. Naik, S.; Malgaresi, G.; You, Z.; Bedrikovetsky, P. Well productivity enhancement by applying nanofluids for wettability alteration. *APPEA J.* **2018**, *58*, 121–129.
38. Divandari, H.; Hemmati-Sarapardeh, A.; Schaffie, M.; Ranjbar, M. Integrating synthesized citric acid-coated magnetite nanoparticles with magnetic fields for enhanced oil recovery: Experimental study and mechanistic understanding. *J. Pet. Sci. Eng.* **2019**, *174*, 425–436. [CrossRef]
39. Moghadasi, R.; Rostami, A.; Hemmati-Sarapardeh, A.; Motie, M. Application of Nanosilica for inhibition of fines migration during low salinity water injection: Experimental study, mechanistic understanding, and model development. *Fuel* **2019**, *242*, 846–862. [CrossRef]
40. Corredor-Rojas, L.M.; Hemmati-Sarapardeh, A.; Husein, M.M.; Dong, M.; Maini, B.B. Rheological behavior of surface modified silica nanoparticles dispersed in partially hydrolyzed polyacrylamide and xanthan gum solutions: Experimental measurements, mechanistic understanding, and model development. *Energy Fuels* **2018**, *32*, 10628–10638. [CrossRef]
41. Shojaati, F.; Riazi, M.; Mousavi, S.H.; Derikvand, Z. Experimental investigation of the inhibitory behavior of metal oxides nanoparticles on asphaltene precipitation. *Colloids Surfaces A Physicochem. Eng. Asp.* **2017**, *531*, 99–110. [CrossRef]
42. Almahfood, M.; Bai, B. The synergistic effects of nanoparticle-surfactant nanofluids in EOR applications. *J. Pet. Sci. Eng.* **2018**, *171*, 196–210. [CrossRef]
43. Bagherpour, S.; Rashidi, A.; Mousavi, S.H.; Izadi, N.; Hamidpour, E. Experimental investigation of carboxylate-alumoxane nanoparticles for the enhanced oil recovery performance. *Colloids Surfaces A Physicochem. Eng. Asp.* **2019**, *563*, 37–49. [CrossRef]
44. Roustaei, A. An evaluation of spontaneous imbibition of water into oil-wet carbonate reservoir cores using nanofluid. *Petrophysics* **2014**, *55*, 31–37.

45. Zhao, M.; Lv, W.; Li, Y.; Dai, C.; Wang, X.; Zhou, H.; Zou, C.; Gao, M.; Zhang, Y.; Wu, Y. Study on the synergy between silica nanoparticles and surfactants for enhanced oil recovery during spontaneous imbibition. *J. Mol. Liq.* **2018**, *261*, 373–378. [CrossRef]

46. Ogolo, N.A.; Olafuyi, O.A.; Onyekonwu, M.O. Enhanced oil recovery using nanoparticles. In Proceedings of the SPE Saudi Arabia Section Technical Symposium and Exhibition, Al-Khobar, Saudi Arabia, 8–11 April 2012; Society of Petroleum Engineers: Richardson, TX, USA.

47. Rezaei, A.; Riazi, M.; Escrochi, M.; Elhaei, R. Integrating surfactant, alkali and nano-fluid flooding for enhanced oil recovery: A mechanistic experimental study of novel chemical combinations. *J. Mol. Liq.* **2020**, 113106. [CrossRef]

48. Suleimanov, B.A.; Ismailov, F.S.; Veliyev, E.F. Nanofluid for enhanced oil recovery. *J. Pet. Sci. Eng.* **2011**, *78*, 431–437. [CrossRef]

49. Hendraningrat, L.; Li, S.; Torsaeter, O. Enhancing oil recovery of low-permeability berea sandstone through optimised nanofluids concentration. In Proceedings of the SPE Enhanced Oil Recovery Conference, Kuala Lumpur, Malaysia, 2–4 July 2013; Society of Petroleum Engineers: Richardson, TX, USA.

50. Asemani, M.; Rabbani, A.R. Detailed FTIR spectroscopy characterization of crude oil extracted asphaltenes: Curve resolve of overlapping bands. *J. Pet. Sci. Eng.* **2020**, *185*, 106618. [CrossRef]

51. Amaefule, J.O.; Altunbay, M.; Tiab, D.; Kersey, D.G.; Keelan, D.K. Enhanced reservoir description: Using core and log data to identify hydraulic (flow) units and predict permeability in uncored intervals/wells. In Proceedings of the SPE Annual Technical Conference and Exhibition, Houston, TX, USA, 3–6 October 1993; Society of Petroleum Engineers: Richardson, TX, USA.

52. Riazi, Z. Application of integrated rock typing and flow units identification methods for an Iranian carbonate reservoir. *J. Pet. Sci. Eng.* **2018**, *160*, 483–497. [CrossRef]

53. Kadkhodaie-Ilkhchi, A.; Kadkhodaie-Ilkhchi, R. A Review of Reservoir Rock Typing Methods in Carbonate Reservoirs: Relation between Geological, Seismic, and Reservoir Rock Types. *Iran. J. Oil Gas. Sci. Technol.* **2018**, *7*, 13–35.

54. Mirzaei-Paiaman, A.; Sabbagh, F.; Ostadhassan, M.; Shafiei, A.; Rezaee, R.; Saboorian-Jooybari, H.; Chene, Z. A further verification of FZI* and PSRTI: Newly developed petrophysical rock typing indices. *J. Pet. Sci. Eng.* **2019**, *175*, 693–705. [CrossRef]

55. Mirzaei-Paiaman, A.; Ostadhassan, M.; Rezaee, R.; Saboorian-Jooybari, H.; Chen, Z. A new approach in petrophysical rock typing. *J. Pet. Sci. Eng.* **2018**, *166*, 445–464. [CrossRef]

56. Moghadasi, R.; Rostami, A.; Hemmati-Sarapardeh, A. Application of nanofluids for treating fines migration during hydraulic fracturing: Experimental study and mechanistic understanding. *Adv. Geo-Energy Res.* **2019**, *3*, 198–206. [CrossRef]

57. Clogston, J.D.; Patri, A.K. Zeta potential measurement. In *Characterization of Nanoparticles Intended for Drug Delivery*; Springer: Berlin/Heidelberg, Germany, 2011; pp. 63–70.

58. Rezvani, H.; Riazi, M.; Tabaei, M.; Kazemzadeh, Y.; Sharifi, M. Experimental investigation of interfacial properties in the EOR mechanisms by the novel synthesized Fe3O4@ Chitosan nanocomposites. *Colloids Surfaces A Physicochem. Eng. Asp.* **2018**, *544*, 15–27. [CrossRef]

59. Idogun, A.K.; Iyagba, E.T.; Ukwotije-Ikwut, R.P.; Aseminaso, A. A review study of oil displacement mechanisms and challenges of nanoparticle enhanced oil recovery. In Proceedings of the SPE Nigeria Annual International Conference and Exhibition, Lagos, Nigeria, 2–4 August 2016; Society of Petroleum Engineers: Richardson, TX, USA.

60. Huang, T.; Han, J.; Agrawal, G.; Sookprasong, P.A. Coupling nanoparticles with waterflooding to increase water sweep efficiency for high fines-containing reservoir-lab and reservoir simulation results. In Proceedings of the SPE Annual Technical Conference and Exhibition, Houston, TX, USA, 28–30 September 2015; Society of Petroleum Engineers: Richardson, TX, USA.

61. You, Z.; Aji, K.; Badalyan, A.; Bedrikovetsky, P. Effect of nanoparticle transport and retention in oilfield rocks on the efficiency of different nanotechnologies in oil industry. In Proceedings of the SPE International Oilfield Nanotechnology Conference and Exhibition, Noordwijk, The Netherlands, 12–14 June 2012; Society of Petroleum Engineers: Richardson, TX, USA.

62. Agi, A.; Junin, R.; Gbadamosi, A. Mechanism governing nanoparticle flow behaviour in porous media: Insight for enhanced oil recovery applications. *Int. Nano Lett.* **2018**, *8*, 49–77. [CrossRef]
63. Sheng, J.J. Enhanced Oil Recovery Field Case Studies: Chapter 19. In *Introduction to MEOR and Its Field Applications in China*; Elsevier Inc.: Amsterdam, The Netherlands, 2013.

MDPI
St. Alban-Anlage 66
4052 Basel
Switzerland
Tel. +41 61 683 77 34
Fax +41 61 302 89 18
www.mdpi.com

Nanomaterials Editorial Office
E-mail: nanomaterials@mdpi.com
www.mdpi.com/journal/nanomaterials